Waste Management Techniques for Improved Environmental and Public Health:

Emerging Research and Opportunities

Sang-Bing Tsai
Zhongshan Institute, University of Electronic Science and Technology of China, China & Wuyi University, China

Zhengxi Yuan
Zhongshan Institute, University of Electronic Science and Technology of China, China

Jian Yu
Research Center for Environment and Sustainable Development of China Civil Aviation, Civil Aviation University of China, China

Xuexin Liu
Capital University of Economics and Business, China

A volume in the Advances in Environmental Engineering and Green Technologies (AEEGT) Book Series

IGI Global
DISSEMINATOR OF KNOWLEDGE

Published in the United States of America by
IGI Global
Engineering Science Reference (an imprint of IGI Global)
701 E. Chocolate Avenue
Hershey PA, USA 17033
Tel: 717-533-8845
Fax: 717-533-8661
E-mail: cust@igi-global.com
Web site: http://www.igi-global.com

Library of Congress Cataloging-in-Publication Data

Names: Tsai, Sang-Bing, 1973- editor.
Title: Waste management techniques for improved environmental and public health : emerging research and opportunities / Sang-Bing Tsai, Zhengxi Yuan, Jian Yu, Xuexin Liu, editors.
Description: Hershey, PA : Engineering Science Reference, 2020. | Includes bibliographical references and index. | Summary: “”This book provides insights into the advancing methods and technologies of reducing pollution and promoting sustainable development. It also examines ecological technologies, risk assessment, and green operation”--Provided by publisher”-- Provided by publisher.
Identifiers: LCCN 2019035509 (print) | LCCN 2019035510 (ebook) | ISBN 9781799819660 (h/c) | ISBN 9781799819677 (s/c) | ISBN 9781799819684 (eISBN)
Subjects: LCSH: Refuse and refuse disposal--Environmental aspects. | Environmental health--Economic aspects. | Sustainable development. | Public health--China. | Green technology.
Classification: LCC TD793 .W38 2020 (print) | LCC TD793 (ebook) | DDC 628.4--dc23
LC record available at https://lccn.loc.gov/2019035509
LC ebook record available at https://lccn.loc.gov/2019035510

This book is published in the IGI Global book series Advances in Environmental Engineering and Green Technologies (AEEGT) (ISSN: 2326-9162; eISSN: 2326-9170)

British Cataloguing in Publication Data
A Cataloguing in Publication record for this book is available from the British Library.

For electronic access to this publication, please contact: eresources@igi-global.com.

Advances in Environmental Engineering and Green Technologies (AEEGT) Book Series

ISSN:2326-9162
EISSN:2326-9170

Editor-in-Chief: Sang-Bing Tsai, University of Electronic Science and Technology of China Zhongshan Institute, China
Ming-Lang Tseng, Lunghwa University of Science and Technology, Taiwan
Yuchi Wang, University of Electronic Science and Technology of China Zhongshan Institute, China

MISSION

Growing awareness and an increased focus on environmental issues such as climate change, energy use, and loss of non-renewable resources have brought about a greater need for research that provides potential solutions to these problems. Research in environmental science and engineering continues to play a vital role in uncovering new opportunities for a "green" future.

The **Advances in Environmental Engineering and Green Technologies (AEEGT)** book series is a mouthpiece for research in all aspects of environmental science, earth science, and green initiatives. This series supports the ongoing research in this field through publishing books that discuss topics within environmental engineering or that deal with the interdisciplinary field of green technologies.

COVERAGE

- Air Quality
- Green Technology
- Radioactive Waste Treatment
- Electric Vehicles
- Cleantech
- Biofilters and Biofiltration
- Renewable Energy
- Pollution Management
- Contaminated Site Remediation
- Green Transportation

IGI Global is currently accepting manuscripts for publication within this series. To submit a proposal for a volume in this series, please contact our Acquisition Editors at Acquisitions@igi-global.com or visit: http://www.igi-global.com/publish/.

The Advances in Environmental Engineering and Green Technologies (AEEGT) Book Series (ISSN 2326-9162) is published by IGI Global, 701 E. Chocolate Avenue, Hershey, PA 17033-1240, USA, www.igi-global.com. This series is composed of titles available for purchase individually; each title is edited to be contextually exclusive from any other title within the series. For pricing and ordering information please visit http://www.igi-global.com/book-series/advances-environmental-engineering-green-technologies/73679. Postmaster: Send all address changes to above address.

Titles in this Series

For a list of additional titles in this series, please visit:
https://www.igi-global.com/book-series/advances-environmental-engineering-green-technologies/73679

Handbook of Research on Resource Management for Pollution and Waste Treatment
Augustine Chioma Affam (University College of Technology Sarawak, Malaysia) and Ezerie Henry Ezechi (Universiti Teknologi Malaysia, Malaysia)
Engineering Science Reference • © 2020 • 795pp • H/C (ISBN: 9781799803690) • US $295.00

Handbook of Research on Globalized Agricultural Trade and New Challenges for Food Security
Vasilii Erokhin (Harbin Engineering University, China) and Tianming Gao (Harbin Engineering University, China)
Engineering Science Reference • © 2020 • 574pp • H/C (ISBN: 9781799810421) • US $285.00

Sustainable Waste Management Challenges in Developing Countries
Agamuthu Pariatamby (Sunway University, Malaysia) Fauziah Shahul Hamid (University of Malaya, Malaysia) and Mehran Sanam Bhatti (University of Malaya, Malaysia)
Engineering Science Reference • © 2020 • 586pp • H/C (ISBN: 9781799801986) • US $225.00

Nano-Phytoremediation Technologies for Groundwater Contaminates Emerging Research and Opportunities
Khushboo Chaudhary (Banasthali University, Tonk Newai, India) Suphiya Khan (Banasthali University, Tonk Newai, India) and Pankaj Kumar Saraswat (ICAR, Krishi Vigyan Kendra, Tamenglong District, Manipur, India)
Engineering Science Reference • © 2020 • 198pp • H/C (ISBN: 9781522590163) • US $165.00

For an entire list of titles in this series, please visit:
https://www.igi-global.com/book-series/advances-environmental-engineering-green-technologies/73679

IGI Global
DISSEMINATOR OF KNOWLEDGE

701 East Chocolate Avenue, Hershey, PA 17033, USA
Tel: 717-533-8845 x100 • Fax: 717-533-8661
E-Mail: cust@igi-global.com • www.igi-global.com

Table of Contents

Preface

The speed of globalization has spurred industry to transfer manufacturing processes filled with toxic substances to regions where wages are lower and environment and labor protection laws are slack. As such, toxic wastes are now being produced in economically backward countries. First, the management of toxic wastes should consider waste reduction during processing. Second, recycling should be implemented in some way. Lastly, the proper handling or disposal of the toxic wastes is of upmost importance. Open-air burning, unsafe recycling, and imperfect recycling equipment have also resulted in various negative effects on communities' public health, including infertility, child development problems, pulmonary function damage, liver and kidney diseases, genetic defects, and mental disorders.

The book's distinctive emphasis is on waste management and improved environmental and public health. The research focuses on waste management, environmental research, environmental pollution, green environment, green technology, sustainability, public health, environment and sustainable development, environment risk assessment, and health risk assessment, emphasizing the use of new theories, technologies, methods, and techniques. The book encompasses theoretical, analytical, and empirical research, comprehensive reviews of relevant research, conceptual frameworks, and case studies of effective applications in this area.

This book contains the following eight chapters.

Chapter 1: An Empirical Study of ISO 14001 Certification and Firm Environmental Performance: ISO 14001 and Firm Environmental Performance

Chapter 2: Big Data, Machine Learning, and Health Knowledge Discovery in the Elderly in China: Health Knowledge Discovery in the Elderly in China

Chapter 3: Urban Solid Waste Management Techniques With Special Reference to Vermicomposting

Chapter 4: Modeling and Prediction of the Zone of Critical Concern: A Solid Waste Management Case of a Higher Education Institute in India Using Probit Regression

Chapter 5: Sustainable Development Model and Innovation of Mineral Resources

Chapter 6: Utilization of Agro-Waste: An Approach for the Administration of a Healthy Environment and Public Domain
Chapter 7: Dairy Waste Treatment and Utilization: A Treatise
Chapter 8: Impact of Trade in Class-A Environmental Products on Economic Growth and Environmental Quality: Simultaneous Equation Analysis in Developing Countries

Acknowledgements are made to IGI Global for professional assistance in the publication of this book as well as to the authors of all the chapters for their contributions.

Sang-Bing Tsai
2019.11.04

Acknowledgment

We acknowledge IGI Global for their professional assistance in the publication of this book, as well as all the chapter authors for their contributions to this book. These editors were supported by University of Electronic Science and Technology of China Zhongshan Institute, China & WUYI University, China & Civil Aviation University of China, China & Environment and Sustainable Development Center of China Civil Aviation University, China & Capital University of Economics and Business, China.

Chapter 1
An Empirical Study of ISO 14001 Certification and Firm Environmental Performance

Yang Wang
Tianjin University, China

Yijun Chen
Tianjin University, China

Qi Xu
Tianjin University, China

Wenqing Wu
Tianjin University, China

ABSTRACT

ISO14001 certification is of great significance to the company's environmental performance. Through the empirical analysis of environmental certification and performance of Listed Companies in China, this chapter draws some valuable conclusions. The initial ISO14001 certification will worsen the environmental performance of the enterprises. The equity nature of the enterprise has a moderating effect on the ISO14001 certification's influence on enterprise environmental performance. The separation rate of the above-mentioned two positions has a moderating effect on the ISO14001 certification's influence on enterprise environmental performance. Based on the above conclusions, this chapter puts forward some reasonable policy suggestions.

DOI: 10.4018/978-1-7998-1966-0.ch001

INTRODUCTION

Currently, enterprises are in the transition period of China's new economy. Under the dual influence of international policies and public opinion, the requirements for the sustainable development and green performance of enterprises are increasing. Using a green development approach in accordance with such international policies will not only help enterprises avoid penalties from the environmental supervision department but also gain these enterprises a series of benefits. Therefore, Chinese enterprises have incorporated the certification of environmental standards such as ISO14001 into their strategies.

In 2017, the total number of global ISO14001 certifications was 319,496, and Asia accounted for 55.7% of these. China ranks first in the world in terms of the total number of ISO14001 certifications, accounting for 36% of the world's total certification. From an industry analysis, it can be seen that ISO-certified companies are basically concentrated in industrial enterprises, and the number of construction industry certifications is at the top.

While the number of ISO14001 certifications by Chinese enterprises is very large, it is still necessary to carefully think about whether environmental performance is actually improved. Although countries in places such as Europe and the Americas have a smaller number of ISO14001 certifications, they have implemented a more stringent environmental standard, namely, EMAS. Compared with ISO14001, EMAS is more concerned with environmental performance, and there are more environmental factors involved. In addition, EMAS sets up specific environmental parameters, while ISO14001 only theoretically gives requirements for environmental standards management systems. To address such defects in ISO14001 (2004), the 2014 ISO/TC officially formulated the "Environmental Management System Requirements and Instructions For Use" as a revision of ISO14001. The new standard puts forward new ideas for strategic environmental management and formalizes the requirements for environmental performance into various provisions, such as 4.4 (Environmental Management System) and 9.1.1 (Monitor, Measurement, Analysis and Evaluation).

In recent years, many scholars have studied the influence of ISO14001 certification on a number of aspects of a company. For instance, Lu Jun and Jiao Shuyan (2011) studied the connection between the paper-making industry and the environmental information disclosure of the construction industry and the connection between environmental performance and financial performance. Yang Dongning and Zhou Changhui (2005) studied enterprises' driving force model for ISO14001 voluntary certification and discussed and analyzed interior driving forces such as the environment orientation of management and external driving forces such as coercive force. Miao Shuyi and other scholars (2014) mainly focused on the important role that ISO14001 certification plays for environmental supervision departments.

There is no definitive answer to whether ISO certification has an impact on environmental performance. Due to the early implementation of ISO14001 in Western countries, the market is freer there, so those regions are more convenient for measuring the impact of ISO14001 certification on environmental performance. Frank Wiengarten (2017) established the MANCOVA model based on the data of 59 Irish factories in 2014 and believed that the environmental performance of ISO-certified companies was significantly higher than that of noncertified companies. Binyam Gr Sediso (2016) investigated 222 residents of three buildings in South Korea that had adopted different environmental standards. Through paper questionnaires on environmental satisfaction, he came to the conclusion that the certified buildings scored higher in terms of lighting and thermal comfort quality than the buildings without certification. Ana Beatriz (2015) studied the relationship between ISO certification and green performance (GP) using models based on quality management (QM), environmental management maturity (EMM), and green supply chain management and using data from 95 Brazilian companies adopting ISO14001 and concluded that there was a strong correlation between green performance and quality management. Deepa Aravind (2011) argues that the implementation of quality standards affects the environmental performance and environmental certification of facilities. Companies with high-quality certification have better environmental performance than their counterparts without certification.

this chapter mainly studies whether ISO has a positive influence on enterprises' environmental performance. Since the ISO14001 system itself provides requirements for management systems, it will definitely affect other aspects, including the assets and strategies of enterprises, through the influence of management decisions. However, it is still unknown whether certification has a positive influence on a company's environmental governance level and governance effect, so the researchers will make the same assessment of the environmental performance of ISO14001. There are different criteria at home and abroad, and determining how to confirm the advantages and disadvantages of enterprises' environmental performance is the main problem that will be solved in this chapter.

BACKGROUND

Theoretical Significance

Although there are many studies on the environmental performance of domestic enterprises, it is unfortunate that these studies generally choose to study the factors affecting the environmental performance of enterprises rather than focusing on the specific influence of ISO14001 certification on environmental performance. this chapter selects a number of representative enterprises as research objects, focusing on the influence of the ISO14001 environmental management system on the economic performance or environmental performance of enterprises and exploring whether there is a positive influence of an enterprise's ISO14001 environmental management system certification on the enterprise's performance or whether there is no significant or even a negative correlation between the certification of the ISO14001 environmental management system and environmental performance.

In addition, the researchers must point out that most of the previous studies on environmental management systems are based on the developed countries, while few scholars have conducted empirical research on developing countries. Environmental performance will be different in places with different institutional backgrounds, and it is affected by various environmental factors. China's market economy has a short development time, so the operation mode of enterprises, the traditional social culture, and the inherent concepts of Chinese people are different from those of Western countries. Therefore, it still needs to be tested whether Western research results are equally applicable to China, given the differing background of China's national condition. At the same time, the report titled *Environmentally Sustainable Future of the People's Republic of China National Environmental Analysis* released on January 14th, 2013 clearly states that among the world's top 10 most air-polluted cities, Chinese cities account for seven of them. China is the world's second largest economy but is also a major polluting country. Therefore, China provides the most profound social background for the academic exploration of the ISO14001 environmental management system.

In summary, on the basis of our country's real conditions, this study takes a number of representative enterprises as empirical research objects. Based on the summarization of a large number of documents, the research proposes the measurement criteria for the organization's daily application and integration implementation level. From the implementation of the standardization to the complete implementation process, the theoretical influence of the ISO14001 environmental management system on the mechanism of enterprise performance is expanded, thus opening the black box of the connection between the ISO14001 environmental management system and enterprise performance. This advancement is not only advantageous to the ever-increasing empirical evidence on the background of China but can also deepen enterprises' understanding and cognition of the implementation of the ISO14001 environmental management system.

Practical Significance

Increasing the ratio of sewage charges to operating income can, in theory, help solve the problem of the inefficient allocation of emission rights, release huge potential for energy conservation and emission reduction, maintain "clean waters and green mountains", and promote economic growth, so it is the best choice for solving environmental problems in China. Therefore, during the implementation of the sewage charge, it is necessary to consider our country's institutions and analyze what the influence of enterprises' decisions to implement ISO14001 certification will be on the environmental performance of enterprises, and the findings are expected to guide the practice of enterprises in the transformation process.

This study hopes to make the enterprise leaders attach more importance to the implementation of the ISO 14001 environmental management system and provide help for the improvement of enterprises' environmental problems by exploring the influence of ISO 14001 certification on enterprises' performance. As part of the "rational person hypothesis", enterprises pursue profit, which is the natural instinct of all social behavior. The researchers speculate that if improving the implementation of environmental management systems can not only improve the economic performance of an enterprise but also improve its environmental performance, then the enterprise will be actively encouraged to implement the ISO14001 environmental management system. Therefore, this study can provide a relatively accurate reference for enterprise leaders to help them improve their "green" image and enhance their enterprise's competitiveness in the market by formulating environmental policies and environmental goals that are appropriate for the enterprises themselves, which has important practical significance for the sustainable development of enterprises.

In conclusion, exploring the internal mechanism of the influence of the ISO environmental management system on enterprise performance is a research effort with theoretical and practical significance.

MAIN FOCUS OF THE CHAPTER[1]

Research Content

On the basis of China's market data and company financial data, the researchers explore and analyze whether the ISO14001 certification of enterprises will affect their environmental performance and explore these relationships and their significance.

(1) Based on the indicator data of enterprises' innovation level and the financial data in the CSMAR database, a mathematical model conforming to the research content is constructed by using the existing measurement settings for specific variables in the reference.

(2) Considering that equity nature will have a regulatory effect on the environmental performance of the ISO14001 certification, the equity nature (state-owned/

private) is introduced as the adjustment variable for regression analysis, and the coefficient of the intersection term is used to judge the adjustment effect and its significance.

(3) Considering that the separation rate of two positions will have a regulatory effect on the environmental performance of the ISO 14001 certification, the separation rate of two positions is introduced into the adjustment variable, and the adjustment function and significance are judged using coefficients such as the intersection term.

Research Method

The first step is to list the virtual variables and set the ISO14001 certification status as a dummy variable. If an enterprise undergoes ISO certification, the variable is set to 1; otherwise, it is set it to 0. According to the data of the database and related financial data online and through a reading of the related reference, evaluation indicators suitable for each variable will be determined. The percentage of enterprise sewage charges in operating income in 2013 to 2017 are selected as a sample for analysis. The ISO certification of enterprises in 2012 to 2016 are used as a sample for analysis. The relevant financial data in 2013 to 2017 are selected as a control variable. The equity nature of the sample and the separation rate of two positions in 2013 to 2017 are used as the regulatory variables. The hypothesis and models are established through analysis and the data for regression analysis are observed.

Document Review and Research Hypothesis

1. The initial certification of the enterprise's ISO14001 status and the enterprise's environmental performance

China's environmental protection system was established relatively late, but based on experience at home and abroad, China's environmental protection system gradually changed from concentration control to total amount control. Since 1978, the sewage charge fee policy has been implemented. By 2018, the sewage charge was officially changed into an environmental protection tax. Is there any correlation between the 40-year sewage charge and China's implementation of ISO14001 environmental protection certification?

Previous scholars have held two views on the influence of ISO certification on the actual environmental performance of enterprises. One view thinks that ISO certification increases the cost of enterprises and reduces the environmental performance of enterprises, while the other view states that ISO certification has improved the environmental management standards of enterprises and will enhance the environmental management performance of enterprises.

Regarding the practical effect of environmental performance, there are the following three points of view.

Raymond's research was conducted from the perspective of practice and adopted an event research method to determine whether enterprises could improve operational efficiency by using management practices under the ISO 14001 standard. The study involved the examination of the performances of certified and noncertified enterprises in the UK and Ireland. The conclusion was that although ISO 14001 has been proven to have a positive and long-term influence on manufacturing cost efficiency and improving the productivity and return on assets of enterprises, there are signs of sample enterprises' operating circles showing diminishing returns in the long term. Ni-Bin believes that the environmental protection problem is one that all developing countries must face when inviting foreign direct investment. However, Ni's study does not show that foreign enterprises must be responsible for the problem because of the high correlation between foreign direct investment and pollution. Ni's study applies unique information from Vietnam's enterprise-level panel data and waste emissions to prove that foreign enterprises are actually more active in acquiring ISO14001 certification. The study provides strong evidence that enterprises' efforts to take charge of social responsibility will ultimately bring benefits to them. Jana believes that eco-innovation has become an important tool to help enterprises reduce environmental influence and gain competitive advantage in an increasingly competitive environment. Jana's study found that competitive interests can regulate and strengthen the connection between ecological innovation and enterprise performance. In addition, the results show that ISO14001 has a positive influence on competitive interests, bringing greater competitive advantages. However, the connection between ISO14001 and enterprise performance has not been found to be statistically significant.

In the past 40 years, ISO14001 certification has been implemented in China. However, as ISO14001 certification only requires the management mode of enterprises, the threshold is not very high. An increasing number of enterprises have obtained ISO14001 certification. The contribution to enterprise environmental performance after ISO14001 certification still needs to be calculated.

There are two methods for measuring environmental performance.

Philippe (2005) studied the impact of ISO14001 on the environmental performance of the paper-making industry, whose environmental performance evaluation indicators of are mainly the volume of pollutant emissions. However, on one point, he is more innovative. That is, the use of the marginal effect to study the influence of an enterprise's pollution emission on the enterprise's economy. Paul (2018) directly used the ESG indicators provided by KLD to measure the environmental performance of an enterprise. ESG indicators mainly include environmental, social, and governmental effects. Since the enterprises included in the KLD database cover most S&P 500 enterprises, excluding Chinese-listed companies, the current paper mainly uses its idea of quantifying environmental performance.

Based on the above documents, this chapter believes that Chinese enterprises will significantly increase their environmental management level after their initial ISO certification and that there will be a great deal of environmental equipment. As a result, for enterprises, the use of stricter environmental monitoring and the improvement of equipment will also improve enterprises' pollution control standards and enable enterprises to pay more governance costs to reach higher standards. Therefore, this chapter predicts that the initial certification of ISO will be consistent with the enterprises' environmental performance.

Based on the above considerations, we propose the following hypothesis:

Hypothesis 1: The enterprise's initial ISO14001 certification will inhibit the enterprise's environmental performance.

2. The moderation effect of equity nature

This chapter argues that since ISO14001 mainly plays a role in the company's management system, the influence of ISO14001 on state-owned and privately owned enterprises' environmental performances is different. This chapter assumes that due to state-owned enterprise management having more concentrated power, the influence on state-owned enterprises is more significant.

According to the research of Tang Guoping (2013) and others, the empirical data on China's A shares indicate the environmental protection investment of Chinese enterprises is mainly done to meet the needs of the government's environmental regulation, and a threshold value of the impact of government regulation exists. Therefore, based on this theory, this chapter studies the impact of the ISO14001 system advocated by the Chinese government on enterprises' environmental performance and investment. The chapter also discusses whether ISO certification has different utility for enterprises with different ownership natures.

This chapter first studies the impact of environmental education on the management maturity of companies. Since the management system of state-owned enterprises has been established for a long time, the management stability and maturity are relatively high. Charbel (2015) believes that environmental education is actually related to the maturity of enterprises' environmental management on the basis of the newly emerged documents regarding enterprises' views of natural resources and green human resource management. Based on this idea, the environmental management maturity of state-owned enterprises is higher, so the environmental education of state-owned enterprises is excellent and the environmental performance is better than that of privately owned enterprises.

Second, the main reason that China charges sewage fees is to solve environmental problems. Environmental problems are typical problems of public goods. This chapter discusses the environmental performance under different property rights systems from the impact of property rights on public goods. Miteva (2019) assesses the role of property rights by applying strict methods similar to experiments and detailed data and takes the ejido public domain as an example. Miteva's article concludes that private property and its impact on the effectiveness of protected areas could reduce forest loss on the Yucatan Peninsula of Mexico in biodiversity hotspots. Therefore, private companies, based on this idea, have better environmental performance.

Third, the internalization of external problems will cause enterprises to include environmental costs in their total costs. Through research, Derk (2009) found that large royalties increase product prices and reduce the demand for final products. Therefore, each patent holder tends to obtain a large amount of royalties at the expense of the sacrifice of competitors. Individual patent holders prefer to negotiate with producers as early as possible and then negotiate with any other rights holders all at the same time. A company with multiple patents wants to talk about each patent in order and at the same time, negotiates payments with other rights holders. In this chapter, the sewage charges belong to the usage rights that enterprises purchase from the government. The government may have a tendency to collect more sewage charges in the short term from the perspective of fiscal revenue and actively negotiate with the enterprises, thus affecting the attitude of the enterprises towards their environmental performance.

Finally, starting from the impact of property rights on enterprise value, this chapter explores whether equity nature has an impact on enterprises and whether enterprise value has a linkage impact on the environmental performance of an enterprise. Thmos (2008) investigates the impact of equity nature on land by applying data from the 2003/04 Cambodian Family Socioeconomic Survey. One document shows that equity nature in rural Cambodia has higher productivity and land value than other plots, and equity nature has little impact on access to credit. The document also investigates whether the introduction of private equity nature has led to a decline in the availability of common property resources. The data provide only weak support for this hypothesis. For the ownership of enterprises' property rights, based on this idea, this chapter assumes that the value of private enterprises is higher than that of state-owned enterprises and that private enterprises' environmental performances are related to the value of those enterprises. Therefore, based on Thmos's work, the environmental performance of private enterprises should be better than state-owned enterprises.

According to the abovementioned line of thought, the environmental performance of private enterprises is higher, but this chapter argues that since ISO14001 mainly plays a role in the company's management system, and the management systems of state-owned enterprises and private enterprises are quite different, the influence of environmental performance is different for state-owned and private enterprises. this chapter assumes that since state-owned enterprises have more centralized management power, the influence on state-owned enterprises is more significant. This is also the innovation point of this chapter.

Based on the above considerations, we propose the following hypothesis:

Hypothesis 2: The nature of the enterprise's equity (state-owned/private) affects the influence of the ISO14001 certification on the enterprise's environmental performance.

3. The moderation effect of the separation rate of the two positions

Stijn Claessens, Simeon Djankov, and Larry H. P Lang (2000) studied the separation rate of two positions and the control of 2,980 companies in nine East Asian countries. They found that in these countries, voting rights often exceed cash flow rights through pyramidal structures and cross-shareholdings. This phenomenon mostly evidenced the separation of ownership and control in family enterprises and small enterprises. More than two-thirds of the studied companies were controlled by a single shareholder. The manager of the share-holding company was often a relative of the controlling shareholder. Older companies were generally controlled by the family, removing the notion that ownership is dispersed over time. Finally, Claessens, Djankov and Lang found that the important enterprise wealth in East Asia is concentrated in a few families. Therefore, the separation rate of the two positions is a kind of enterprise governance method that is often adopted by enterprises.

Johan J. Graafland (2018) suggests that the use of ISO14001 certification is a symbol that aims to improve the public image of an enterprise and has no substantial beneficial influence on the ecological environment. Graafland's study explores scientific documents and argues that previous studies have neglected possible mediators through which ISO14001 might have a positive and indirect influence on ecology. In addition, the study examines the indirect influence of ISO 14001 certification on the ecological outcomes of 3633 small enterprises in 12 European countries through the application of structural equation modeling. The findings prove that the influence of ISO 14001 on the ecological outcomes of small enterprises can be mediated through supply chain cooperation or collaboration with training institutions or local organizations. On the basis of these results, it can be concluded that promoting ISO14001 certification in small enterprises is particularly useful if it is combined with external networks involved in facilitating implementation.

Rafael La Porta, Florencio Lopez-de-Silanes, Andrei Shleifer, and Robert W. Vishny (2000) studied the influence of the interaction between the ultimate ownership structure and the external governance environment on the dividend policy of agricultural listed companies from the dual perspectives of the state and the region. The study mainly found that at the level of internal ultimate ownership, the separation of the two controlling shares of the ultimate controlling shareholder has a negative impact on the cash dividend policy of agricultural listed companies. The higher the separation of the two rights of cash flow and control rights is, the lower the willingness to pay and the level of payment will be. The external governance environment is positively related to the cash dividend policy; that is, the better the external governance environment is, the higher the willingness and level of cash dividends paid by agricultural listed enterprises will be. The improvement of the external governance environment will help remove the negative influence of the separation of the ultimate controlling shareholder's two controlling interests on the willingness of agricultural listed companies to pay cash dividends. Additionally, there is no significant connection between the intersection of the two controlling interests and the cash dividend payment level, namely, external governance. The enhancement of the environment can effectively prohibit the behaviors of the ultimate controlling shareholder from infringing on the power of small and medium shareholders for the individual's interests and increase the willingness to pay cash dividends.

However, there are also different opinions. Yu Cong and Martin Freedman (2011) studied the connection between the good governance structure of a company and both environmental performance and the disclosure of that performance. The study chose a US company that emits toxic pollutants as a sample for research to measure pollution performance through the application of a method that includes both emission toxicology and community population density. The main performance indicators were verified on the basis of the company governance structure and pollution disclosure assessments presented in the accounting documents. The study results indicated that there is no clear link between good governance structure and improved environmental performance. This may indicate that reports on pollution performance are better than the real results.

Since management leaders of an enterprise emphasize the short-term interests of the enterprise while the owner of the enterprise pays more attention to the long-term value of the enterprise, this chapter contends that whether or not the owner is the management leader of the company will have a significant influence on the company's environmental performance. If the separation rate of the two positions is low, business owners will pay more attention to short-term economic benefits; therefore, this chapter assumes that ISO 14001 has a negative influence on environmental performance and that enterprise owners are unwilling to sacrifice profits of a certain year for the certification. However, for companies with a high separation rate, enterprise owners will place more emphasis on long-term value and the company's image in order to maintain stock prices, so ISO 14001 certification will have a positive effect on environmental performance.

Based on the above considerations, we propose the following hypothesis:

Hypothesis 3: The separation ratio of the two positions will have a regulatory effect on whether ISO14001 certification affects environmental performance.

RESEARCH DESIGN

Data Processing Brief

The data chosen in this empirical process came from the CSMAR database and the Wind database. In addition, when checking whether the enterprise is undergoing ISO14001 certification, the study confirmed the certification condition on reliable websites such as the company's official website to ensure the reliability of the research data, thus leading to a reliable study conclusion. The original sample collected in the research was 300 GEM enterprises (2010-2017). The samples that had partly missing accounting data and samples that were unable to be located were ruled out, so the final number came to 122 samples. The annual observations of 122 samples (including explanatory variables, interpretation variables, control variables, and adjustment variables) were included in the Stata panel data, leading to 560 observations. The study also performed WINSOR tailing processing on the data before the subsequent empirical regression analysis to eliminate the interference of outliers on the research results.

Research Variables and Variable Data Sources

1. Measurement of environmental performance

This chapter selected the proportion of the company's sewage charges to operating income as a measurement indicator of environmental performance. This chapter believe that the better the enterprise environmental performance is, the lower the proportion of sewage charges to operating income will be since sewage charges are generally levied on the concentration of enterprise pollutants, so a higher payment means that the company emits more pollutants.

The financial data of sewage charges and operating income come from the CSMAR database.

2. Independent variables - whether ISO14001 certification has been performed

A virtual variable was introduced to indicate whether the enterprise has passed ISO14001 certification. If it has, the variable is set as 1, otherwise, it is set as 0.

3. Control variables and settings

In order to study whether the enterprise having passed ISO14001 certification has an influence on its environmental performance, control variables must be added to the model, and the influence of other variables on the explained variables should be excluded as much as possible to better study the influence of the dependent variables. Considering the influence of the company's scale, number of years in operation, and financial status variables on environmental performance, the following control variables are set:

a. Company scale: The total assets of a company can reflect the scale of the company. To make the statistical analysis of the data clearer, the logarithm of the total assets is used to indicate the scale of the company.
b. Whether the enterprise's ISO14001 certification can promote its environmental performance.
c. Rate of investment profits: An indicator that measures how much net profit is generated per unit of investment to reflect the company's profitability, which may affect its environmental performance.
d. Business operation index: This is used to measure the business status and cash flow of a company.

The data of the above control variables are all from the CSMAR database.

4. Adjust variables and settings
    a. Equity nature

Using the classification standard of Guotaian, the 122 sample enterprises were divided into state-owned, privately owned and foreign enterprises on the basis of the equity nature of the enterprises, and the virtual variables were set. The data on the nature of enterprise equity are all from the CSMAR database.

b. Separation rate of two positions

The separation of the two positions is the separation theory of the enterprises' ownership and control (management rights). Chandler believes that the intensification of shareholding and the specialization of management make managers who have specialized management knowledge and monopolize business information actually hold the control of an enterprise, leading to the "separation of the two rights." The ratio of the separation of ownership and management rights reflects, to some extent, how much the CEO controls the company, which may have further influence on environmental performance.

Each variable's definition and measurement method in this study are shown in the following table:

Models and Data

Panel data are used to study and analyze the influence of ISO 14001 certification on enterprise environmental performance; taking multiple sections in time series and at these interfaces, sample-related financial values are simultaneously selected to form matrix sample data. The panel data have the two dimensions of time series and section, which form a plane that is significantly different from the basic one-dimensional data. The panel data can more reasonably reflect the connection between

Table1. Variables' definitions and measurements

Type	Variable Name	Variable representation	Measurement
Dependent variable	Camp ratio	Y	Sewage charges / operating income
Explanatory variable	ISO14001 certification	ISO	Pass: 1, not pass: 0
Control variable	Enterprise scale	Size	Logarithm of total assets
	Enterprise operating years	Firmage	Current year - establishment year
	Investment profit rate	ROI	Annual profit before tax / total investment
	Operation index	Ope	Operating cash flow / operating cash
Moderating variable	Equity nature	Holder	State-owned: 1,private or foreign: 0
	Separation rate of the two positions	Sep	Separation ratio of business ownership and management rights

sample variables and take the time factor into consideration, which can, on one hand, reflect the connection between sample and variables at a certain period and, on the other hand, take into account the law of how the sample changes with time.

Based on the previous hypothesis and the selected type of research for the data analysis, the following model was established:

Model 1: Study how the initial ISO14001 certification inhibits the company's environmental performance, i.e., verification of hypothesis 1:

$$Y_{1it} = \beta_0 + \beta_1 Size + \beta_2 Age + \beta_3 ROI + \beta_4 Ope + \varepsilon_{it} \quad (1)$$

Model 2: Study how the equity nature affects the influence of ISO14001 certification on environmental performance, i.e., verification of hypothesis 2:

$$Y_{2it} = \gamma_0 + \gamma_1 ISO + \gamma_2 Size + \gamma_3 Age + \gamma_4 ROI + \gamma_5 Ope + \gamma_6 HolISO + \varepsilon_{it} \quad (2)$$

Model 3: Study how the separation rate of two rights influence the environmental performance of ISO14001 certification, i.e., verification of hypothesis 3:

$$Y_{3it} = \eta_0 + \eta_1 ISO + \eta_2 Size + \eta_3 Age + \eta_4 ROI + \eta_5 Ope + \eta_6 HoLISO\varepsilon_{it} + \varepsilon_{it} \quad (3)$$

In the above, i is the number of cross-section observations, i.e., the number of samples, i = 1, 2, 3, ..., 122, and t is the time series, i.e., the statistical year, t = 2013, 2011, ..., 2017. The dependent variable Y is the ratio of sewage charges to operating income, and the independent variable ISO is ISO14001, which is set as 1 if the enterprise has the certification, otherwise, it is set as 0. After multiple tests, the regression results of ISO variable with a lag of three years were the most obvious, so the lag variable of three years was selected in this chapter; the control variables include size (enterprise ε_{it} scale), age (enterprise operating years), ROI (return on investment rate), and Ope (operation index); the moderating variables include Holder (equity nature) and Sep (separation rate of two positions), which represents the random variable.

This study was conducted using Stata's panel data test method, adopting a mixed effect model, random effect model and fixed effect model for analysis. To test the reliability of the model, the F-Test was used to judge the advantages and disadvantages of the hybrid model and the fixed model. The Hausman Test was used to compare the advantages and disadvantages of the random model and the fixed model. The study ultimately shows that the fixed effect model is optimal.

In the final phase of the empirical research, the key variable substitution method in the model was selected to test the robustness of the model, and the ROI (return on investment) model was replaced by ROA (return on assets) and EPS (EPS), which were conducted in the research; finally, the robustness of the model was verified.

EMPIRICAL ANALYSIS AND TESTING

Sample Description and Statistical Analysis

It is worth noting that the explained variables and explanatory variables can potentially vary with time and individual changes. The differences between the variables of the sample selected in this study that are due to individual differences are called differences between groups; the differences in time are called differences within groups.

The estimator between groups takes a time average for each individual explanatory variable and then uses the average to make a regression, while the estimator within the group takes the time average of each individual explanatory variable in the individual fixed-effects model and then removes the fixed effect. In this process, each explanatory variable removes the individual average information and uses only the estimated amount of the dispersion information within the group individually. The overall estimator is used to estimate the overall parameters, taking into consideration the time factor and the factors of individual differences. Therefore, this empirical study uses more overall estimation statistics for analysis and explanation. The operating income per share is differs greatly. The differences between the groups are more obvious than the differences within the groups, but that is mainly in terms of the differences in earnings per share of the listed companies.

The average value proportion of ISO14001-certified companies is 0.3464286, which indicates that most companies do not have ISO14001 certification; this reflects the fact that the number of environmental certifications of listed companies in China is generally not high.

There are few differences between the groups and within groups regarding company scale. It can be seen that there are random phenomena both in time and in individual explanatory variables.

The number of years a business has been in operation is quite different, but the listed companies are basically operation-stable companies. Therefore, it makes sense that the average operation time of listed companies in China is 17.91 years.

The individual differences in the ROI of listed companies in China are greater than the time differences. The overall ROI of the time is relatively average, but the average ROI of listed companies is 3.9%, which is the same as the yield of the five-

Table 2. Descriptive statistics

Name	Sample number		Average	Standard deviation	Minimum	Maximum
EP1	Overall	568	1.764	2.865	0.001	30.722
	Between groups	112		2.577	0.009	21.026
	Within the group	5		1.271	-5.264	13.417
IS0	Overall	560	0.346	0.476	0	1.000
	Between groups	112		0.427	0	1.000
	Within the group	5		0212	-0.453	1.1464
Size	Overall	560	9.864	0.562	8.501	11.409
	Between groups	112		0.548	8.698	11.381
	Within the group	5		0133	9.040	10.638
Age	Overall	560	17.910	4.308	6.000	31.000
	Between groups	112		4.083	8.000	29.000
	Within the group	5		1415	15.910	19.910
ROI	Overall	560	0.039	0.098	-1.058	0.837
	Between groups	112		0.059	-.1858	0.377
	Within the group	5		0078	-0.833	0.498
Ope	Overall	560	0.696	4.013	-49.978	72.229
	Between groups	112		1.822	-9.164	15.204
	Within the group	5		3579	-40.117	57.721
Sep	Overall	558	6.425	8.487	0	36.334
	Between groups	112		8.030	0	24.269
	Within the group	4.982		2.814	-6.936	22.855
Holder	Overall	560	0.242	0.429	0	1.000
	Between groups	112		0.399	0	1.000
	Within the group	5		0160	-.557	0.842

year government bond but with larger risk, reflecting that China's listed companies do not create a good investment income.

The larger the variance between the ISO and the separation ratio of the two positions, the greater the time effect over the individual effect.

The separation ratio of the two rights in the group is smaller than the separation rate between the groups. The overall separation rate of the two rights in listed companies in China is expected to be 6.4, which is in accordance with the general phenomenon.

There are large differences between the groups of ISO and equity nature due to the differences in the equity of each company.

Sample Correlation Analysis

The paper first conducted descriptive statistics and then checked the kurtosis and skewness of each variable data before the correlation analysis and regression analysis in order to test the reliability of the model. If the kurtosis is large and whether there is left or right deviation, an appropriate range was chosen to perform the WINSOR tailing processing to eliminate the effects of unreasonably extreme values.

It can be seen from the table that the correlation matrix between the independent variable, the control variable and the regulatory variable was less than 0.5, with the majority being less than 0.1, so the interference resulting from the multicollinearity model and the research conclusion could be excluded again.

It is not difficult to draw some possible conclusions from the positives and negatives of the correlation, e.g., the larger the enterprises are, the lower the earnings per share are. When a company is first created, due to limited funding sources and limited personnel, the company is relatively small in scale, with less financing and fewer total shares. However, as enterprises develop and their financing methods expand after listing, enterprises have more funds to develop and the shares are then further diluted, so the earnings per share decreases. The longer the business is in operation, the larger the company scale is, which is in accordance with objective law. The performance forecast of state-owned enterprises is significantly related to ISO14001 certification, company scale, operation, etc., probably because state-owned enterprises' operations are relatively stable and such enterprises face fewer uncertainties, so the effect is more significant in predicting enterprise performance.

Table 3. Statistical Correlation Analysis

Variables	EP1	IS03	size	Age	ROI	OPE	SEP	HOLISO
EP1	1.0000							
IS03	0.0426	1						
size	-0.1508**	0.1482**	1					
Age	-0.0450	0.0456	0.1280***	1				
ROI	0.0418	-0.0038	0.1344	0.0406	1			
OPE	0.0151	0.0408	-0.0179	-0.0238	-0.0007	1		
SEP	-0.1462**	0.1689***	0.032**	0.0403	0.0445	0.0144	1	
HOLISO	-0.0261	0.7779***	0.1989	0.1249***	-0.0296	0.0443	0.1336***	1

Panel Data Analysis

1. Panel Data Model

This study needs to select the most reliable model form among the mixed effect model, the random effects model and the fixed effects model.

First, the F-Test is used to test the individual effect in order to select the mixed or fixed effect model. Then, the Hausman Test is used to test the correlation between the individual effect and the explanatory variable in order to decide whether to use the random effect model or the fixed effect model. The specific test values are shown in Table 4, Table 5, and Table 6.

2 Regression analysis Results

The panel data were input into Stata, then the above model and related test methods were used for regression analysis. Finally, the results are shown in Tables 4 and 5:

Table 4. Regression analysis and test of model 1 and model 2

	Model 1		
	Pooled	**RE**	**FE**
ISO	0.404	0.937***	1.098***
Size	-0.802***	-1.215***	1.600***
Firmage	-0.020	-0.035	-0.040
ROI	1.313	-0.037	-0.172
Ope	0.006	0.003	-0.003
_cons	9.828***	14.057***	17.897***
Industry	included	included	included
	0.04	0.04	0.06
Observations	122	122	122
F-Test			16.95***
Hausman-test			20.34***

P.S: *p<10%, **p<5%, ***p<1%, N=122

Table 5 Regression analysis and test of model 3 and model 4

	Model 2			Model 3		
	Pooled	RE	FE	Pooled	RE	FE
ISO	1.978***	1.702***	1.981***	1.676	1.482***	1.639***
ISO×Holder	-0.117	-1.288***	1.517***	-	-	-
ISO×Sep	-	-	-	-0.802***	-0.700***	-0.071***
Size	-0.762***	-1.188***	-1.636***	1.454	-1.237***	-1.634***
Firmage	0.014	-0.032	-0.038	-0.044	-0.038	-0.044
ROI	1.359	-0.056	-0.180	-0.044	-0.046	-0.165
Ope	0.015	-0.003	-0.003	-0.002	-0.002	-0.002
_cons	9.421***	13.790***	18.284***	12.24***	14.337***	18.323***
Industry	included	included	included	included	included	included
	0.03	0.07	0.07	0.05	0.07	0.07
Observations	122	122	122	122	122	122
F-Test			17.17			17.00**
Hausman-test			21.46***			20.09**

P.S: *p<10%, **p<5%, ***p<1%, N=122

SOLUTIONS AND RECOMMENDATIONS

Description and Analysis of Regression Results

Among the four models above, it can be determined via an F-Test that compared to the mixed effect model, the fixed effect model is better. The Hausman test shows that compared to the random effect model, the fixed effect model is still better. The best model for this study is the fixed effect model.

Therefore, using the data of the above four fixed effect models, we verified the three hypotheses and drew related conclusions:

a. When model 1 is below the 5% significance level, hypothesis 1 cannot be rejected. The coefficient of Init is 0.740, indicating that when the other conditions are unchanged, the initial ISO14001 certification will reduce the environmental performance of the enterprises and enable the enterprise to spend more on sewage charges.
b. Through the observation of model 2, it can be concluded that the intersection of ISO and equity nature is very significant, meaning that the state-owned or non-state-owned equity nature has a regulatory effect on the implementation of ISO14001 certification to promote the environmental performance of

enterprises; that is, when the result is under the 1% level of significance, hypothesis 2 cannot be rejected. Both ISO and Holder are virtual variables, and the coefficients of the intersection can reflect that the state-owned enterprises earned ISO14001 certification in the sample year. The coefficient of the balance item is negative, indicating that the environmental performance of the state-owned enterprises was better than that of the private enterprises, and the state-owned enterprises often spent a smaller amount on sewage charges, which means that when the nature of the equity is the state-owned enterprise, there is an environmental performance enhancement effect. In addition, when the enterprise share was privately held, the intersection was 0, and the coefficient of the CEO category was still positive, indicating that in the private enterprise, ISO14001 certification will reduce the environmental performance of enterprises and enable the enterprises to spend more on sewage charges.

c. It is not difficult to determine, through the observation of model 3, that the intersection of ISO and the separation rate of two rights is very significant, which means that the separation rate of two rights has a regulatory effect on the environmental performance of enterprises. Therefore, when under the 1% significance level, hypothesis 3 cannot be rejected. Observing the coefficient of the intersection term -0.071 as a negative number, it can be seen that in a two-right separation enterprise, the ISO14001 certification will improve the environmental performance of the enterprise and reduce the sewage cost of the enterprise. In a two-right-in-one enterprise, the coefficient of ISO is 1.639, so ISO14001 certification will reduce the environmental performance of the enterprise.

Robustness Test

To test the empirical results' reliability and to ensure that the results of the regression are stable rather than contingent upon the sample analysis, the robustness of the model needs to be tested. This chapter chose the method of substitution variables for robust testing.

The researchers introduced indicators that could also represent the profitability of the company—ROA (return on assets) or EPS (EPS) instead of ROI (return on investment) as one of the control variables—and brought them into four models for research and analysis.

For the robustness test, the following two variables were introduced:

From the above results, it can be seen that after replacing the ROI with the ROA, the ISO14001 certification situation (initial year and annual), that is, both the coefficient of positive or negative and the significance of the independent variable of the study, did not change much and could still sufficiently verify the four hypotheses

Table 6. Table of variables introduced

Type	Variable name	Variable representation	Test method
Control variable	Asset profit rate	ROA	Net profit / total assets
Control variable	Interest per share	EPS	After-tax profit / total share capital

Table 7. Regression results after substituting key variables

Variable	Model 1	Model 2	Model 3
ISO	3.74***	4.57***	4.70***
Size	-3.35***	-3.45***	-3.45***
Age	-0.83	-0.79	-0.92
ROA	-0.54	-0.55	-0.47
Ope	-0.22	-0.23	-0.14
HolISO	-	-2.74**	-
SepISO	-	-	-2.83***
_cons	4.02***	4.13***	4.14***
Observations	560	560	560

P.S: *p<10%, **p<5%, ***p<1%, N=122

Table 8 Regression results after substituting key variables

Variable	Model 1	Model 2	Model 3
ISO	3.74***	4.58***	4.71***
Size	-3.32***	-3.41***	-3.44***
Age	-0.85	-0.81	-0.95
EPS	-0.08	-0.20	0.17
Ope	-0.19	-0.20	-0.09
HolISO	-	-2.75***	-
SepISO	-	-	-2.84**
_cons	3.98***	4.08***	4.13***
Observations	560	560	560

P.S: *p<10%, **p<5%, ***p<1%, N=122

of this chapter. The adjustment effect of the equity nature was verified again in the new model, and its coefficient was still negative, and the significance level did not change. Under the 1% significance level, hypothesis 2 cannot be rejected. The adjustment effect of the two-right separation rate of the enterprise was again verified. There was no change in the positive and negative coefficients or the significance level, and under the 1% significance level, hypothesis 3 cannot be rejected.

Based on the above analysis, the research model is very robust, and the operation results of Stata are reliable; based on reliable results, reliable conclusions can be drawn.

CONCLUSION AND REVELATIONS

Based on the domestic GEM from 122 listed companies and using empirical research, this chapter used panel data analysis and related models to verify that analysis. On the basis of the existing documentary summary and reference, the study can draw some conclusions from the empirical analysis of the influence of enterprises' ISO14001 certification on environmental performance as a reference for enterprises to strengthen their operational management and for the government to enhance enterprises' environmental performance and environmental protection.

At the time of the research, this chapter separately explored the obvious differences of the influence of the ISO14001 certification on the environmental performance when the certification was delayed for one year, two years and three years. Finally, it was found that ISO14001 certification only affected the environmental performance when it was delayed by one year. After the initial ISO14001 environment certification, the company's sewage charges would increase, and the environmental performance would also improve.

In addition, the research in this chapter indicates that except that for the first year, ISO14001 certification has a restraining effect on an enterprise's environmental performance; the certification of ISO14001 in any other year will cause enterprises to spend more on sewage charges. This phenomenon may be explained by the following reasons, which were collected through existing documents and materials: enterprises need to pay a certain cost when applying for the ISO14001 certification, and enterprises need to invest a great deal of money in the early, middle and late stages of certification, including in upgrading the company's environmental protection facilities to meet certification standards or seeking reimbursement for certification procedures and the attendance of certified personnel. As a result, enterprises will increase their production scale to compensate for the cost of the environmental certification in the previous year. During the process of increasing production, enterprises may emit more pollutants. On the other hand, enterprise supervision under the ISO14001 certification is also more stringent. These factors will lead to the environmental performance of the company being worse in the following year of ISO14001 certification.

As for the adjustment effect of the nature of equity (state-owned/private) on the environmental performance of ISO14001 certification, this chapter draws the conclusion that among enterprises with ISO14001 certification, the innovation performance of state-owned enterprises is higher than that of private enterprises, and state-owned enterprises often spend a smaller amount on sewage charges. This is easy to understand since state-owned companies are generally large in scale, there are high-level talents and top international environmental protection equipment, and their funds are sufficient. Therefore, ISO14001-certified state-owned enterprises will not blindly increase their output and thus increase pollutant emissions in the attempt to control pollutant emissions, and their innovation performance will inevitably outperform that of private enterprises. ISO14001 certification will worsen the environmental performance of private enterprises.

The separation of the two positions and agency costs will inevitably occur in every enterprise that is in the process of development. Therefore, the researchers included the separation rate of the two rights as a regulatory variable in the research. ISO14001 certification will worsen the environmental performance of two-right-in-one enterprise, while it will improve the environmental performance of enterprises with high separation rates. This can be verified using the research conclusions of some documents and materials that show that, in the process of an enterprise's continuous development, the founding CEO may lack relevant organization management skills; as a result, hiring a professional CEO for business management can improve the management level and enhance the performance of an enterprise. Therefore, in the two-rights separation enterprises, when enterprises conduct ISO14001 certification, the professional CEO will make relatively correct and efficient strategic adjustments to avoid blindly adopting radical marketing and other measures to improve corporate financial performance and maintain high environmental performance.

In general, in order to improve their competitiveness and company image, attract consumers, and respond to the country's requirements for environmental protection, when applying for ISO14001 certification, enterprises will expand their scale to make up for the cost, which will lead to an increase in sewage charges. Ordinary scaled, private enterprises do not own enough environmental protection equipment to deal with the pollutants; therefore, they will have a high amount of sewage charges, which will worsen their environmental performance. Therefore, they must seek a balance between the benefits and the cost of sewage charges in order to maximize the benefits of the company. At the same time, such enterprises should try to consider the long-term benefits rather than short-term profit, and they should update their environmental protection equipment in a timely manner. Enterprises should actively understand the country's policies regarding environmental pollutant emissions, try their best to minimize pollutants or optimize the links that can be discharged for levying pollutants, and improve their environmental performance.

As for the environmental protection department, the standards of ISO14001 certification should be improved, and the cost of certification should be reduced, as this would help lead enterprises to realize the importance of environmental protection. At the same time, the environmental protection department should continuously conduct inspections of enterprises that have passed ISO14001 certification to reduce the phenomenon of enterprises blindly expanding in order to make up for the cost of certification. On the other hand, certified enterprises should be given some preferential policies and strategic support, and environmental experts should be sent to these enterprises to provide guidance on reducing pollutant emissions so that enterprises can spend less on sewage charges and enhance their environmental performance.

With the increasingly improved status of environmental protection in social and economic development, all enterprises and environmental departments should strengthen their environmental management and hire relevant professional talents. This will bring long-term benefits that cannot be ignored to the environment and related institutions.

REFERENCES

Aravind, D., & Christmann, P. (2011). Decoupling of standard implementation from certification: Does quality of iso 14001 implementation affect facilities' environmental performance? *Business Ethics Quarterly*, *21*(1), 73–102. doi:10.5840/beq20112114

Barla, P. (2006). ISO 14001 certification and environmental performance in Quebec's pulp and paper industry. *Journal of Environmental Economics and Management*, *53*(3), 291–306. doi:10.1016/j.jeem.2006.10.004

Boiral, O., & Sala, J.-M. (1998). Environmental management: Should industry adopt ISO 14001? *Business Horizons, Elsevier*, *41*(1), 57–64. doi:10.1016/S0007-6813(98)90065-9

Claessens, S., Djankov, S., & Lang, L. H. (2000). The separation of ownership and control in East Asian corporations. *Journal of Financial Economics*, *58*(1-2), 81–112. doi:10.1016/S0304-405X(00)00067-2

Cong, Y., & Freedman, M. (2011). Corporate governance and environmental performance and disclosures. *Advances in Accounting*, *27*(2), 223–232. doi:10.1016/j.adiac.2011.05.005

Darnall, N., & Sides, S. (2008). Assessing the Performance of Voluntary Environmental Programs: Does Certification Matter? *Policy Studies Journal: the Journal of the Policy Studies Organization*, *36*(1), 95–117. doi:10.1111/j.1541-0072.2007.00255.x

Griffin, P. A., Neururer, T., & Sun, E. Y. (2018). Environmental performance and analyst information processing costs. *Journal of Corporate Finance*. doi:10.1016/j.jcorpfin.2018.08.008

Jabbour, C., de Freitas, T., Soubihia, D., Gunasekaran, A., & Jabbour, A. (2015). Green and competitive: Empirical evidence from ISO 9001 certified Brazilian companies. *The TQM Journal*, *27*(1), 22–41. doi:10.1108/TQM-01-2013-0013

Jabbour, C. J. C. (2015). Environmental training and environmental management maturity of brazilian companies with iso14001: Empirical evidence. *Journal of Cleaner Production*, *96*, 331–338. doi:10.1016/j.jclepro.2013.10.039

Jun, L., & Jiao, S. (2011). An Empirical Study on the Relationship between Environmental Disclosure, Environmental Performance and Financial Performance. *Journal of Shanxi University of Finance and Economics*, *33*(1), 109–116.

La Porta, R., Lopez-de-Silanes, F., Shleifer, A., & Vishny, R. W. (2000). Agency problems and dividend policies around the world. *The Journal of Finance*, *55*(1), 1–33. doi:10.1111/0022-1082.00199

Miao, S., Ji, D., Guo, H., Wang, W., & Wang, M. (2014). Application of ISO14001 Environmental Management System Certification in Environmental Monitoring Department. *Environment and Sustainable Development*, *39*(06), 132–134.

Miteva, D. A., Ellis, P. W., Ellis, E. A., & Griscom, B. W. (2019). The role of property rights in shaping the effectiveness of protected areas and resisting forest loss in the Yucatan Peninsula. *PLoS One*, *14*(5), e0215820. doi:10.1371/journal.pone.0215820 PMID:31067246

Ni, B., Tamechika, H., Otsuki, T., & Honda, K. (2019). Does ISO14001 raise firms' awareness of environmental protection? *The case of Vietnam. Environment and Development Economics*, *24*(1), 47–66. doi:10.1017/S1355770X18000396

Sediso, B. G., & Lee, M. S. (2016). Indoor environmental quality in Korean green building certification criteria—certified office buildings—occupant satisfaction and performance. *Science and Technology for the Built Environment*, *22*(5), 606–618. doi:10.1080/23744731.2016.1176849

Tang, G., & Li, L. (2013). Study on the Structure and Distribution Characteristics of Enterprise Environmental Protection Investment——Evidence from A-share Listed Companies from 2008 to 2011. *Audit and Economic Research*, *28*(04), 94–103.

Wiengarten, F., Humphreys, P., Onofrei, G., & Fynes, B. (2017). The adoption of multiple certification standards: perceived performance implications of quality, environmental and health & safety certifications. *Production Planning and Control, TBC*(2), 131-141.

Yang, D., & Zhou, C. (2005). The driving force of enterprises adopting standardized environmental management system voluntarily: Theoretical framework and empirical analysis. *Management World*, (02): 85–95.

Zobel, T. (2013). ISO 14001 certification in manufacturing firms: A tool for those in need or an indication of greenness? *Journal of Cleaner Production*, 43.

ADDITIONAL READING

Aravind, D., & Christmann, P. (2011). Decoupling of standard implementation from certification: Does quality of ISO 14001 implementation affect facilities' environmental performance? *Business Ethics Quarterly*, *21*(1), 73–102. doi:10.5840/beq20112114

Barla, P. (2007). ISO 14001 certification and environmental performance in Quebec's pulp and paper industry. *Journal of Environmental Economics and Management*, *53*(3), 291–306. doi:10.1016/j.jeem.2006.10.004

Boiral, O., & Henri, J. F. (2012). Modelling the impact of ISO 14001 on environmental performance: A comparative approach. *Journal of Environmental Management*, *99*, 84–97. doi:10.1016/j.jenvman.2012.01.007 PMID:22326757

Clark, D. J., & Pereau, J. C. (2009). Fragmented property rights and royalty bargaining. *Journal of Economic Behavior & Organization*, *72*(1), 546–553. doi:10.1016/j.jebo.2009.05.022

Heras-Saizarbitoria, I., & Boiral, O. (2013). ISO 9001 and ISO 14001: Towards a research agenda on management system standards. *International Journal of Management Reviews*, *15*(1), 47–65. doi:10.1111/j.1468-2370.2012.00334.x

Kitazawa, S., & Sarkis, J. (2000). The relationship between ISO 14001 and continuous source reduction programs. *International Journal of Operations & Production Management*, *20*(2), 225–248. doi:10.1108/01443570010304279

McGuire, W. (2014). The effect of ISO 14001 on environmental regulatory compliance in China. *Ecological Economics, 105*, 254–264. doi:10.1016/j.ecolecon.2014.06.007

Treacy, R., Humphreys, P., McIvor, R., & Lo, C. (2019). ISO14001 certification and operating performance: A practice-based view. *International Journal of Production Economics, 208*, 319–328. doi:10.1016/j.ijpe.2018.12.012

Zobel, T. (2013). ISO 14001 certification in manufacturing firms: A tool for those in need or an indication of greenness? *Journal of Cleaner Production, 43*, 37–44. doi:10.1016/j.jclepro.2012.12.014

KEY TERMS AND DEFINITIONS

Environmental Performance: Enterprise environmental performance is the result of an enterprise's environmental management. This chapter quantifies the environmental performance with the amount of annual sewage charge.

Equity Nature: Equity nature is traditionally divided into state shares, state-owned legal person shares, social legal person shares, natural person shares. This chapter simplifies it into state holding and private holding.

Hausman Test: The Hausmann test can be used to normalize the OLS and WLS estimates to see if the difference exceeds the sampling error range.

ISO14001 Certification: ISO 14001 is the code name for environmental management system certification. The ISO 14000 series of standards is an environmental management system standard developed by the International Organization for Standardization. It addresses global environmental pollution and ecological damage, as well as major environmental problems such as ozone depletion, global warming and the loss of biodiversity.

Moderating Variable: The relationship between dependent variable and independent variable is a function of moderating variable and is affected by moderating variable.

Separation Rate of Two Positions: The difference between control and ownership. The calculation results are obtained according to the equity control chain.

Sewage Charge: According to the laws and regulations of a country, polluters should pay a certain amount of money to the state according to the type, quantity and concentration of pollutants they discharge.

Chapter 2
Big Data, Machine Learning, and Health Knowledge Discovery in the Elderly in China

Bin Ding
Xi'an Jiaotong-Liverpool University, China

Dongxiao Gu
Hefei University of Technology, China

Zheng Jiang
Hefei University of Technology, China

ABSTRACT

According to the national strategic plan for healthy aging and the construction of the pension system in China, it is expected that by 2020 the population of elderly aged 60 and above will reach 255 million, accounting for about 17.8% of the total population. Currently, population aging is a serious social problem in China, and thus, health status of the elderly becomes increasingly critical. The present research uses machine learning to identify factors influencing elderly's health status and life satisfaction with data from the Chinese Longitudinal Healthy Longevity Survey. The results show that some common factors are important for both self-rated health status and life satisfaction for elderly, namely positive and optimistic attitudes, a healthy diet, and economic status. Health status and life satisfaction also have their unique predicting factors, such as mobility ability for health status and living conditions for life satisfaction. Theoretical and practical implications of the findings are discussed.

DOI: 10.4018/978-1-7998-1966-0.ch002

INTRODUCTION

According to the "Deep research and investment strategy report of 2017-2022 China's pension industry" released by Zhiyan (2017), the population of people aged 60 and above amounted to 240 million in 2017, accounting for 17.3% of the total population in China. The population of people aged 65 and above reached 158.31 million, accounting for 11.4% of the total population in China. By 2020, the population of people aged 60 and above will reach 255 million; the elderly who live alone or in "empty nests" will be about 118 million; dependency ratio of elderly will rise to about 28% (Lu & Wang, 2017). Aging issue in China is becoming more serious and visible, manifested in disability, high age, and empty nests in the old populaiton. The "13th Five-Year Plan" of the State Council has established strategic plans for aging and pension system, aiming at improving the social security system and the pension system, so that the living, spiritual and social participation aspects of elderly will be guaranteed. Against this backdrop, increasing research attention has been devoted to health issue of the old population.

China is rapidly entering a population aging society, and population aging process in China is faster than other low-income and middle-income countries. With the sharp increase of old population, great pressure and challenges have been brought on current social security and pension system. A pressing issue for now is to gain a scientific understanding of the health status of the elderly under current economic conditions in China. Albeit the medical advances and increment of people's living condition make people's average lifespan increased dramatically, with the increase of age, the physiological function of elderly would inevitably decline to some extent. They may have cognitive and physiological dysfunction, chronic diseases, decreased mental states, and so forth, and hence they would rely heavily on social welfare such as medical care, healthcare system, community service, and nursing services. Therefore, to gain a deeper understanding of the key factors affecting the health status of the elderly is imperative. This research aims to address this research inquiry.

Early research predominantly studies elderly health from a medical and biological perspective (Perez, Bauerl, & Collado, 2014; Goldman et al., 2005; De Martinis, Sirufo, & Ginaldi, 2017). As research progresses, scholars investigate factors affecting elderly's health from the perspectives of sociology and economics. Research shows that people's living environment and their personal conditions has profound impact on aging, and these factors start to influence the aging process from a very early stage and last for a prolong time (WHO, 2015). For example, health status of the elderly could be affected by social environment in which people live including family, neighborhood and community environment (e.g., Kang, Kim, & lee, 2008). Meanwhile, sufficient supporting service, such as community services and healthcare system, plays an important role in affecting elderly's health and improving their life quality (e.g., Hosseini & Bahraminejad, 2014). Research further reveals that personal factors including socioeconomic status, physical activities, eating habits, and psychological conditions can help reduce chronic disease risk and thus improve physical and mental health of elderly (Bae et al., 2010; Kwon & Kim, 2016; Yu, Kang, & Kim, 2015).

Prior research on elderly's health predominantly used traditional statistical analytic methods such as ordinary least square regression. Because traditional analytic methods normally focus on a limited number of theory-driven predictors, some hidden knowledge regarding the influencing factors of health in elderly cannot be found. This analytic drawback impedes us from gaining a comprehensive and holist understanding of a given data. The rapid development of machine learning technology has brought new opportunities for knowledge discovery in social science. Machine learning technology can discover hidden knowledge or patterns in large-scale datasets, and subsequently provide important information for interpretation and prediction. With the development of algorithms, machine learning technology has been widely used in the fields of finance, medicine, transportation, and business and management (Zhu, Xie, Wang & Yan, 2017; Obermeyer & Emanuel, 2016; Jahangiri & Rakha, 2015; Bose & Mahapatra, 2001). Against this backdrop, we aim to use machine learning technology to investigate factors affecting the health status and life satisfaction of elderly. By doing so, we endeavor to potentially explore new knowledge of health issue in elderly, and provide empirical evidence for scientific advancement and policy making.

RELATED RESEARCH

Health in Elderly

The concept of healthy aging was first proposed at the World Health Assembly in 1987, and it was then adopted by the WHO as a developmental strategy to address population aging at the World Conference on Aging in 1990. In order to actively respond to the aging of population and steadily improve the health of the old population, the WHO released the Global Report on Aging and Health in October 2015 (WHO, 2015). The main goal of healthy aging is to improve life quality of the elderly, shorten the survival period of the disease, and prolong the life expectancy. Similarly, China has also formulated and issued a series of policies to guide the development of healthy aging. For example, in March 2017, China established the "13th Five-Year Plan for Healthy Aging", indicating that healthy aging has become an important part in China's macro strategic plan.

Existing research on the health of the elderly reveals that macro-level strategies, such as retirement system, pension, community service and support, medical system, usage of medical information system may reduce the incidence of diseases, extend life expectancy, improve life quality, and health status (Jiang, 2016; Zhong, 2013; Wang et al., 2016; Liang, et al., 2017; Liu & Liu, 2018; Gu et al., 2019). Moreover, micro-level factors such as socioeconomic status, personal habits, demographics, marriage status, participating in social activities, living with others or not, air pollution, and eating habits play important roles in influencing psychical and psychological health of elderly with different samples in multiple countries (Xue & Ge, 2017; Li et al., 2018; Jin, Zhou & Zhai, 2017; Ali et al., 2018; Choi et al., 2015; Bind et al., 2016; Bahler et al., 2015; Giacalone et al., 2016).

Machine Learning and Social Science Knowledge Discovery

Social science includes those subjects that study human behaviors and societal phenomena. For example, traditional social science include subjects such as sociology, political science, economics, and social psychology. Recently, cross-disciplinary subjects such as social computing and information science are counted as social science as well, because these subjects combine the theories of traditional social sciences with advanced analytic techniques. For example, data mining, especially machine learning, has been used to investigate social phenomena with large scale datasets. Machine learning techniques include a series of computer-intensive methods designed to discover patterns and hidden knowledge in data. It uses methods such as response/adaptive design, data processing, non-responsive adjustment and weighting, and data modeling. Using machine learning techniques, social scientists have discovered very interesting results which go above and beyond those revealed with traditional analytic methods such as regression. For example, data mining has been used for discovering factors influencing customer satisfaction in banks, for decreasing project risks in automobile industry, for predicting student success in the future, for improving education and teaching quality and for predicting participants' chronic diseases (Ali et al., 2018; Choi et al., 2015; Bind et al., 2016; Giacalone et al., 2016).

Next, we present our research aims and research questions, dataset, analytic methodology and results of this research.

Research Aims

The release of large-scale health datasets offers significant value for research advancement, for example, China general social survey (CGSS), Chinese Family Panel Studies (CFPS), China Health and Retirement Longitudinal Study (CHARLS),

China Education Panel Survey (CEPS), China Household Finance Survey (CHFS), etc. The China Longitudinal Health and Longevity Survey (CLHLS) is one of these open access datasets that has been used widely. CLHLS is supported by several research institutes, such as NIA/NIH, China Social Sciences Foundation, Duke University, Max Planck Institute and others. The study started in 1998, and the data collection has covered 23 provinces and more than 860 counties/districts in China. The CLHLS dataset and follow-up datasets have been used in a large number of research studies, and many interesting findings have been revealed for improving healthy aging and active aging in China (e.g., Shi et al., 2015; Gao et al., 2017; Zeng et al., 2018). The current research also uses the CLHLS dataset and aims to contribute to the healthy aging literature. More specifically, we are interested in exploring macro- and micro-factors that influence old people's health and life satisfaction in the elderly using the CLHLS dataset with data mining-based knowledge discovery methods. We aim to offer a ranking of the importance of the predicting factors for elderly's health and life satisfaction. By gaining a comprehensive and holist understanding of healthy aging in China, we are able to move forward the scientific research in elderly health, and offer practical implication and suggestions for policy making and strategy formulation.

METHOD

Dataset

We used the latest dataset of CLHLS in 2014. The dataset was accessed from Peking University's Open Research Data platform (http://opendata.pku.edu.cn/dataverse/CHADS). Researchers of the CLHLS conducted face-to-face interviews with individuals using internationally compatible questionnaires. As the CLHLS project aims to investigate which factors out of a large set of social, behavioral, biological, and environmental factors, play an important role in healthy longevity, survey questions covered a wide range of factors including health, disability, demographic, family, socioeconomic, community, and behavioral risk factors for mortality and healthy longevity for elderly. Table 1 shows sample questions in the survey. Full questionnaire can be found at http://opendata.pku.edu.cn/dataverse/CHADS.

The 2014 CLHLS dataset we used in this research contained a total of 7,192 subjects and 886 variables. Before we started data analysis, we first conducted a preliminary examination and data cleaning of the dataset. We only kept the variables that were potentially related to the present research, and we deleted those subjects with empty inputs. This deletion process resulted in a final dataset containing 6,577 subjects and 206 variables. It should be noted that the CLHLS dataset included extensive data on a population of the oldest-old aged 80-112 and younger elders aged 65-79. We thought that people with an age over 105 might have difficulty in participating in the survey and cooperating with the data collection procedure. Therefore, we decided to delete those subjects with age over 105. The final dataset contained 6,314 subjects with 206 variables.

Variable Selection and Coding

Before we present the dataset and analysis in detail, we would like to discuss the definition and measurement of health in elderly. World Health Organization (WHO) defines health as "a state of complete physical, mental and social well-being and not merely the absence of disease or infirmity". This definition indicates that health is a complex construct with rich content, and it should not be measured with a single indicator. In the CLHLS dataset, we chose two variables to measure elderly's health—self-rated health and life satisfaction. Self-rated health reflected the general physiological condition of a person. It was measured as "How do you rate your health at present?" Because life satisfaction reflected an overall assessment of people's life quality, much existing research used life satisfaction to measure health status (e.g., Li, Feng, & Yang, 2014). Therefore, we used life satisfaction as another proxy for elderly's health. In CLHLS, life satisfaction was measured by asking the following question: "How do you rate your life at present?"

Table 1. Examples of questions in the Chinese People's Health and Longevity Influencing Factors Survey

D11 Do you now perform the following activities regularly? (please choose one from frequency on the right)	**1 almost every day.**	**2 not every day, but at least once a week.**	**3 not every week, but at least once a month.**	**4 not every month, but sometimes.**	**5 never.**
D11.1 Housework (cooking, taking care of kids)	1	2	3	4	5
D11.2 Personal outdoor activities	1	2	3	4	5
D11.3 Garden work	1	2	3	4	5
D11.4 Read newspapers/books	1	2	3	4	5
D11.5 Raise domestic animals	1	2	3	4	5
D11.6 Play cards and/or mah-jong	1	2	3	4	5
D11.7 Watch TV and/or listen to radio	1	2	3	4	5

In machine learning, health of the elderly can be considered as a classification problem, that is, according to their health status, elderly are classified into different levels or groups, and then identify the predicting factors for health in elderly. More specifically, we classify elderly with good and bad health condition, and find their similarities and differences by effectively distinguishing the two groups of elderly. We used permutation importance and Shapley value of Shapley Additive Explanations (Shap) technology to identify the most important explanatory variables, and used Gradient Classifier (GBDT) to predict self-rated health and life satisfaction.

RESULTS

Feature Selection Results of Self-Rated Health

Data were analyzed with an integrated feature selection algorithm. We regarded those variables whose means were above the average as important predicting factor for our outcome variables, and used those variables for further feature selection. The feature selection procedure yielded 44 features for self-rated health.

Based on the GBDT model, we examined the respective effect of the 44 factors on self-rated health when examined alone. Table 2 shows the permutation importance of the 44 factors. Higher ranking in the table indicated larger effect of the features on self-rated health. The features at the top had larger effects than the features at the bottom. The effects were shown in the form of mean plus/minus standard deviation.

We thought that self-rated health, change of health and life satisfaction all reflected health status of the respondents. It was not out of expectation that the three of them had large correlation with one another. Results in Table 2 showed that the top five factors influencing elderly's self-rated health status were sleeping quality, whether having limitation in activities, looking at the bright side of things, age, and cost of outpatient. Quality of sleeping ranked as a top predictor for self-rated health. As elderly's age increased, it was common that they had decreasing sleeping quality and sleeping time. Having high quality sleeping played a significant role in influencing elderly's self-rated health. Having been limited in activities was ranked as another top factor influencing elderly's self-rated health. Mobility restrictions are signs of physical health problems in elderly, encountering such restrictions greatly affected elderly's range of activities as well as their psychological conditions. Our results indicated that looking at bright side of things helped people stay optimistic when dealing with all kinds of things in life, and furthermore, enabled people to stay psychological healthy at an old age. Moreover, aging played a significant role in influencing elderly's self-rated health, indicating that as people get older, their physical and physiological conditions would inevitably decrease and thus affected

Table 2. The permutation importance of self_rated health

Ranking	Effect	Feature
1	0.0660 ± 0.0050	life_quality_selfeval
2	0.0541 ± 0.0036	health_change_selfeval
3	0.0105 ± 0.0065	quality_sleep
4	0.0089 ± 0.0021	health_activity_limit
5	0.0067 ± 0.0042	feel_positive
6	0.0061 ± 0.0035	trueage
7	0.0052 ± 0.0022	cost_outpatient
8	0.0041 ± 0.0016	crouch_stand
9	0.0035 ± 0.0023	feel_useless
10	0.0032 ± 0.0021	happy_as_young
11	0.0032 ± 0.0009	suffering_heart
12	0.0030 ± 0.0018	pick_up
13	0.0030 ± 0.0012	num_teeth
14	0.0030 ± 0.0018	walk_kilo
15	0.0029 ± 0.0009	suffering_lung
16	0.0028 ± 0.0023	province
17	0.0025 ± 0.0037	feel_neat_clean
18	0.0025 ± 0.0015	feel_bad
19	0.0025 ± 0.0028	feel_fear_anxiety
20	0.0024 ± 0.0016	direct_height
21	0.002186±0.0008	hearing_difficult
22	0.001965±0.0009	finan_compare
23	0.001521±0.0012	times_sick
24	0.001489±0.0004	carry_weight
25	0.001426±0.0006	mother_alive
26	0.001331±0.0006	suffering_cvd
27	0.001077±0.0010	fruit_frequency
28	0.001077±0.0006	money_from_child
29	0.001077±0.0008	age_born_first_child
30	0.000951±0.0009	time_sleep
31	0.000856±0.0009	egg_often_now
32	0.000856±0.0006	house_destroyed
33	0.000856±0.0006	gar_often_ago
34	0.000729±0.0004	saltveg_often_ago

continued on following page

Table 2. Continued

Ranking	Effect	Feature
35	0.000602±0.0005	drink_now
36	0.000602±0.0003	grow_vegetables
37	0.000475±0.0004	nut_milk_now
38	0.000412±0.0007	height_indirect
39	0.000317±0.0010	feel_lonely
40	0.000317±0.0009	exercise_now
41	0.000158±0.0007	suga_often_now
42	0.000158±0.0000	suffering_diabets
43	−0.000063±0.0002	suffering_arthritis
44	−0.000095±0.0006	income_family

their health status. Annual cost on outpatients signified a decrease in health status as reported by participants themselves.

Among all the 44 factors, factors influencing elderly's self-rated health status roughly fell into the following categories. The first category was chronic diseases. Suffering heart disease, stroke and cerebrovascular, lung disease such as bronchitis, emphysema and asthma, arthritis, and diabetes were the most notable chronic diseases that affected elderly's health. Second, mobility abilities such as being able to crouch and stand up continuously for three times, to pick up a book from the floor, to walk continuously for one kilometer by themselves, to lift a weight of 5kg, to perform garden work, and to do exercise regularly were also decisive for elderly to remain healthy. Moreover, our results showed that having good psychological conditions especially having positive emotions were helpful to make elderly physically healthy. Top ranked emotion and mood related questions included looking at the bright side of things, as happy as they were younger, feeling fearful or anxious, and feeling lonely and isolated. We also found that having a healthy diet impacted elderly's health condition, including the frequency of eating fruit, eggs, vegetables, sugar, garlic and nut. Finally, financial condition including economic status compared with others in the local area, how much money they received from their children, and income per capita of the household were found to be important factors for elderly's self-rated health.

Besides the influencing factors mentioned above, some of other features interested us as well, for example, height of the elderly and province where they were born. Past research showed that as people become older, their height would shrink to some extent, which served as a good indicator of elderly's physiological condition, which in turn, signified people's health status (Huang et al., 2013). This finding offered a possibly simple way to keep good track of the health condition of elderly. This result was particularly interesting because past studies using the CLHLS dataset seldom revealed height as an indicator of elderly's health. We hope our finding encouraged more interesting results on height as a factor to predicting elderly's health condition. In addition, geography locations such as the provinces in which participants were born also influenced their health. Our explanation would be that where participants were born might influence their eating habits and life style, which were important for people for staying healthy.

The GBDT model enabled us to rank the importance of factors affecting self-rated health condition of the elderly when examined alone. Next, we used a prediction analysis method, namely the shap value interpretation method, to analyze and visualize how each factor influenced elderly's health when examined together. We plotted the shap values for each feature as shown in Figure 1. The figure was generated by sorting the sum of the shap value of each feature, starting from the most influential features to the least. The horizontal axis was the shap value, which represented the extent to the factors affected the outcome variable, i,e., self-rated health. Each point in a feature represented a sample. The color of the point represented high or low of the feature, wherein red meant high value while blue meant low value. It can be seen from the figure that the predictive ability of the features was consistent with the importance ranking in Table 2. The overall predicting ability of the features from high to low was ranked from top to bottom.

Figure 1 shows that the shap results largely echoes with the GBDT results as shown in Table 2. Sleeping quality, looking at the bright side of things, whether having limitation in activities, and cost of outpatient were still ranked as top factors. The only difference was that feeling useless was ranked as top five when using shap, just above age. As discussed above, feeling useless reflected a low level of psychological health condition of elderly, indicating that it was crucial to stay psychologically positive for elderly. Negative emotion sometimes hurt more than positive emotion helps.

Knowledge Discovery for Life Satisfaction

Following a similar procedure, we analyzed how factors affected life satisfaction of the elderly using the GBDT integration model. We selected the variables whose means were above the average, ultimately resulting in 47 features when predicting life satisfaction. Table 3 presented the permutation importance of the 47 factors. The features were ranked according to their importance for life satisfaction from high to low. The features at the top had larger effects on life satisfaction than the features at the bottom.

Figure 1. The shap value summary plot of self_rated health

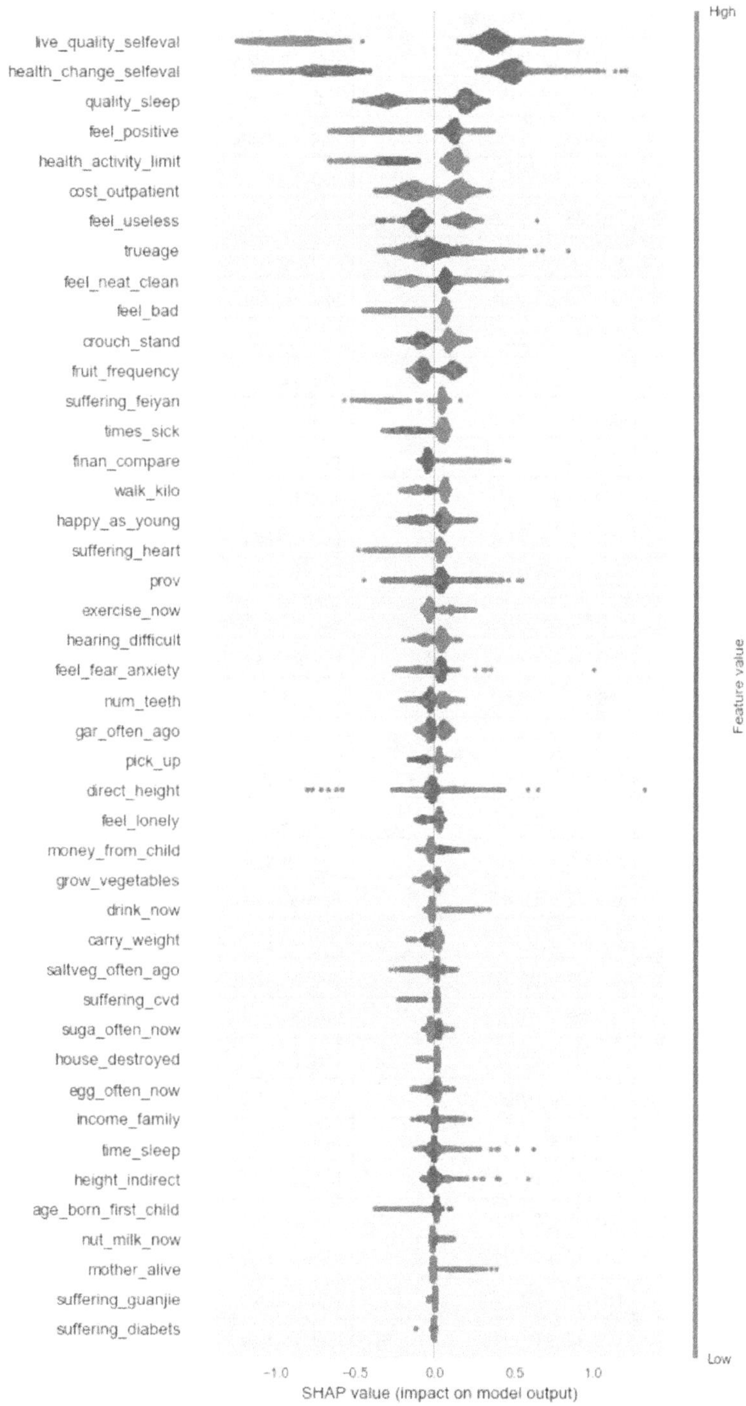

Table 3. The permutation importance of life satisfaction

Rank	Weight	Feature
1	0.0502 ±0.0022	live_health_selfeval
2	0.0161 ±0.0050	finan_compare
3	0.0103 ±0.0052	feel_neat_clean
4	0.0087 ±0.0023	happy_as_young
5	0.0083 ±0.0020	prov
6	0.0070 ±0.0029	feel_positive
7	0.0066 ±0.0040	income_family
8	0.0059 ±0.0014	cost_outpatient
9	0.0049 ±0.0036	trueage
10	0.0048 ±0.0015	waist
11	0.0045 ±0.0028	money_from_childr
12	0.0042 ±0.0021	rate_heart
13	0.0042 ±0.0041	house_elder_year
14	0.0042 ±0.0023	weight
15	0.0042 ±0.0014	age_first_child
16	0.0041 ±0.0008	flavor_have
17	0.0031 ±0.0003	tv
18	0.0031 ±0.0015	main_food
19	0.0029 ±0.0008	fish_often_now
20	0.0028 ±0.0028	finan_enough
21	0.0028 ±0.0019	age_born_first_child
22	0.0028 ±0.0015	health_activity_limit
23	0.0028 ±0.0006	mushroom_now
24	0.0027 ± 0.0023	calf_quguang
25	0.0022 ± 0.0017	health_change_selfeval
26	0.0022 ± 0.0024	feel_fear_anxiety
27	0.0022 ± 0.0008	hosue_elder_type
28	0.0021 ± 0.0012	quality_sleep
29	0.0020 ± 0.0029	height_indirect
30	0.0019 ± 0.0019	direct_height
31	0.0018 ± 0.0023	source_water_now
32	0.0018 ± 0.0010	feel_lonely
33	0.0017 ± 0.0024	feel_useless
34	0.0014 ± 0.0008	milk_often_ago

continued on following page

Table 3. Continued

Rank	Weight	Feature
35	0.0012 ± 0.0016	source_water_child
36	0.0012 ± 0.0016	height_indirect_1
37	0.0012 ± 0.0012	vita_now
38	0.0012 ± 0.0023	house_destroyed
39	0.0011 ± 0.0014	fruit_frequency
40	0.0011 ± 0.0015	commu_psy_avalib
41	0.0011 ± 0.0017	saltveg_often_now
42	0.0010 ± 0.0021	public_car_self
43	0.0009 ± 0.0008	decision_own
44	0.0006 ± 0.0016	commu_social_avalib
45	0.0006 ± 0.0022	live_with_family
46	0.0001 ± 0.0016	milk_often_now
47	−0.0002±0.0014	house_smelled

As found in the previous section, self-rated health and life satisfaction were closely related to each other as dependent variables. Except for self-rated health, the top five important factors for life satisfaction were economic status compared with others in the local area, keeping belongings neat and clean, feeling as happy as when they were younger, province where they were born, and looking at the bright side of things. High economic status could satisfy material and medical needs of elderly, ensuring a relatively high quality of life. Keeping belongings neat and clean ranked as the second most important factor for life satisfaction, indicating that it was necessary and important to have a clean and tidy living environment to make elderly happy about their life. It was also very important to be optimistic and stay happy for elderly. Having a good mood helped people to be psychologically healthy, which in turn, made people happy about their life in general. In which province elderly were born was ranked top when influencing elderly's life satisfaction. This result was similar to self-rated health, and province had even larger effect on life satisfaction than self-rated health.

Factors influencing elderly's life satisfaction roughly fell into the following categories. First, economic situation played a significant role in making elderly satisfied with their life. Economic factors included economic status compared to others as appeared in top five, income per capita of the household, money from children, and whether they have enough financial support. Enough financial support could ensure that elderly's daily and urgent needs were satisfied, so that they could have a better life, ultimately they would be happy about their life. Second, a better psychological status made elderly have higher life satisfaction, for example, feeling as happy as when they were younger, looking at the bright side of things, feeling fearful and anxious, and feeling lonely and useless. Psychological health was not only related to physically health, it also made elderly happy about their life. Third, our results showed that a healthy diet impacted elderly's health, in addition, it affected their satisfaction with their life as well. The intake of fish, mushroom or algae, milk, vitamin, and fruit, flavor they mainly have, type of staple food, and salt-preserved vegetables greatly impacted their life satisfaction. Fourth, we also found that living condition played a role in determining elderly's life satisfaction. Living condition included whether they would like to keep things neat and clean, the year when house was build, type of dwelling, whether there were water or dampness in the house from broken pipes, leaks, heavy rain or floods, whether there was frequently a mildew odor or musty smell in the house, community environment such as what kind of social services were available in the community and whether community provided psychological consulting service. Furthermore, some physical examination indicators, such as waist circumference, weight, heart rate, height, height of calf also affected elderly's life satisfaction. These results have not been fully explored in past research using the CLHLS dataset, thus providing some interesting insights for explaining elderly's life satisfaction. Finally, health-related factors also affected elderly's life satisfaction. Factors such as cost spent on outpatient during the last year, limitation in activities due to health problems, and self-rated health change, might make elderly worry about their health condition, which in turn affect their general happiness with their life.

Next, we used shap to obtain the global ranking of the features and visualized how the features affected elderly's life satisfaction. As shown in Figure 2, the results were essentially the same as those obtained using GBDT, except that the order of the features obtained using shap was a bit different from those using GBDT. For example, whether respondents had enough financial support was ranked top eighth using shap but the 20th using GBDT, and house damage due to dampness and flood was also ranked top using shap.

DISCUSSION

In this research, we investigate the factors influencing elderly's self-rated health status and life satisfaction using the China Longitudinal Health and Longevity Survey 2014 dataset. We find that some common factors appear to be important for both self-rated health status and life satisfaction, namely positive and optimistic attitudes, a healthy diet, and economic status. Results show that people with more positive and optimistic attitudes, frequently eating nutrition such as milk, eggs, fruit

Figure 2. The shap value summary plot of life satisfaction

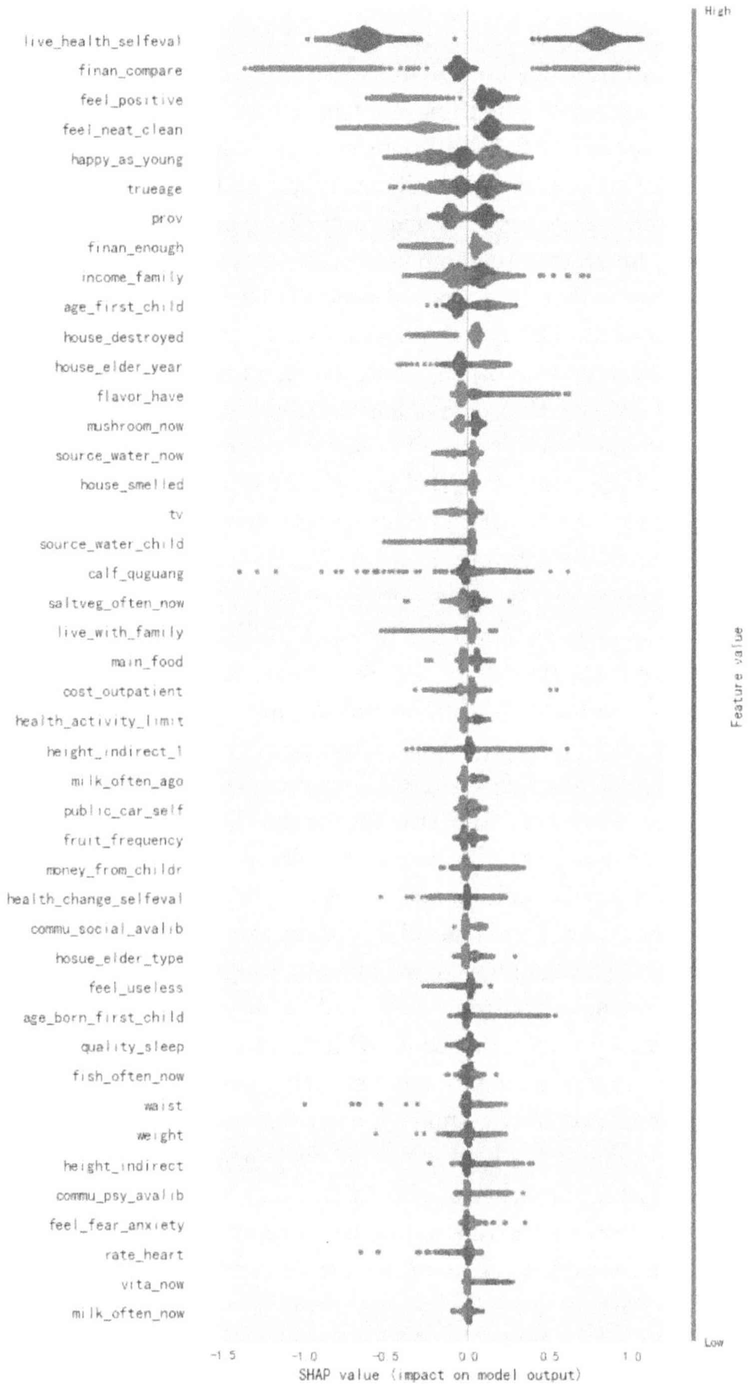

and vegetables, and with sufficient financial support are more likely to have better self-rated health condition and are more satisfied with their life. These findings echo with past research that psychological resilience and optimism are positively related to self-rated health and self-rated successful aging (e.g., Hardy, Concato, & Gill, 2002; Wagnild, 2003; Steptoe, Wright, Kunz-Ebrecht, & Iliffe, 2006). More relatedly, Zeng and colleagues use the CLHLS 2008 dataset with a sample of 16,566 and find that psychological resilience greatly contributes to longevity at all ages and the results become more pronounced when age increases (Zeng et al., 2017). Moreover, Shen and Zeng (2010) reveal that psychological resilience reduces mortality risk among all elderly aged more than 65 years old using CLHLS longitudinal data (Shen & Zeng, 2011). Our results also echo with past research and demonstrate a correlation between eating habits and self-rated health and life satisfaction. For example, Shi et al (2015) use the 1998 CLHLS dataset and find that daily vegetable and fruit intake is associated with low mortality risk, but the intake of self-preserved vegetables is associated with an approximately 10% increased risk of death (Shi, Zhang, Byles, Martin, Avery, & Taylor, 2015). Food habits of older people can affect their nutritional intake and thus affect their health. Rolls et al. (2015) examine the neural basis of age-related differences in taste and food acceptability, and find that age differences in the activation of different tastes are related to differences in the pleasure levels of food in certain brain regions (Rolls, Kellerhals, & Nichols, 2015). This work provides the basis for understanding the underlying neural basis of differences in food acceptability between age groups, and suggests that the acceptability of food and beverages is related to age. Therefore, it is necessary to develop specific recipe for old people, or to adjust the diet structure of the elderly, thereby ensuring basic nutritional intake and health of the elderly. Similar as what we find, much research has shown that financial condition influences elderly's psychological and physical health. Research shows that financial constrain exerts deleterious influence on elderly's psychological health and well-being (Borgm Hallberg, & Blomqvist, 2006). Because financial constrain may result in an increase in depression, such depression potentially leads to a decreased level of satisfaction with life. Lack of financial support may also lead to delay in seeing doctors and ultimately worsen the disease. Income inequality is shown found to have a negative effect on elderly's well-being and an indirect effect on elderly's health, which echoes with what we find using the CLHLS dataset (Hu &Lu, 2012).

Our research contributes to research on healthy and active aging in several ways. First, psychological health plays an important role in predicting elderly's health and life satisfaction. Positive attitudes and optimism help elderly to maintain healthy bodies. It is crucial to establish or improve related community service, nursing service, and so forth to ensure a positive psychological state for elderly. It is also crucial for younger generation, children of the elderly population, especially those who live alone or with empty nests, to realize the importance for elderly to stay psychological healthy. Spending more time with the old people, going backing home more often, calling them as many as possible could potentially prevent them from falling into long-term loneliness or anxiety, or even depression.

Second, we find that economic status of the elderly has a noticeable impact on health status and life satisfaction. The elderly population has a decline in body function and is prone to a variety of chronic diseases, which greatly require long-term healthcare services, frequent visits and hospitalizations. High medical expenses on outpatient or hospitalization are heavy burden for old people with poor economic conditions. Improving the healthcare and pension system would be a long-term investment. Meanwhile, offering more community healthcare service stations to elderly could potentially enhance the accessibility of health services and make outpatient lighter, cheaper, and easier for elderly. Community healthcare service stations can also better track medical history and take action in a timely manner, so that elderly could avoid a large amount of consumption on acute care and potentially reduce the financial burden.

Third, because of aging, chewing and digesting systems of the elderly may become weaker, affecting their daily diet and nutrient intake, and thus threatening their health. Even worse, some oldest-old people have problems with swallowing, so that foods need careful chewing may pose a greater risk to them. Our results show that some food such as vegetables, fruits, eggs, milk, etc. are correlated with elderly's health condition and life satisfaction. Therefore, it is necessary to ensure the intake of these kinds of food and prepare according dietary structure for elderly. Government and health institutes should broadcast related knowledge to the public and draw more attention to necessary nutrition intake for elderly.

Fourth, there are some unique factors influencing each of the two outcome variables. For instance, chronic diseases and mobility abilities greatly affect elderly's self-rated health. It is thus necessary to put more effort on preventing and healing chronic diseases for elderly. Strengthening the health care system to provide knowledge on chronic diseases for people and offer sufficient medical support in community or hospitals for the public would be very important. It would also be necessary to stress the importance of doing light exercise for elderly to maintain mobility abilities in the process of aging. Physical examination indicators, living condition, and health-related indicators are found to be important for keeping elderly satisfied with their life. Past research shows that weight and waist-to-abdominal ratios are potential predictors for obesity and may influence heart function and lead to hypertension (Yan et al., 2018), offering possible explanations on why some physical exanimation indicators play important roles in influencing elderly's self-rated health, it is thus important to keep such indicators in a normal range. Because some simple physiological indicators may have good predictive power on the health of the elderly, we could establish a stratified system for physical examination of the elderly, so that light and fast examinations could be done in the community more frequently, and comprehensive physical examination could be done yearly. Therefore, a track record and early warning of the physiological health status of the elderly would be helpful to prevent and track illness. Our results on living condition show that good living environment makes elderly happy about their life. A dry, clean and convenient living environment may keep elderly away from some disease such as rheumatism and ensure that elderly see a doctor in a timely manner. Therefore, it is necessary to ensure a clean, neat and comfortable living environment in terms of both house condition and community service for elderly.

Finally, our findings indicate the possibility and superiority of using machine learning analytic technique for revealing hidden knowledge in large-scale dataset. Our results are largely consistent with existing findings using the CLHLS dataset. In addition, we reveal some interesting results that are not found before, such as the effect of province. The use of machine learning technique sheds new light on the CLHLS dataset, and moreover, it could further advance the social science by exploring more hidden knowledge without offering hypotheses ex ante. Using machine learning also enables us to have a global understanding of the CLHLS dataset and health condition of elderly in China. Based on our findings, government could make potential strategic adjustment on the healthcare system to facilitate healthy and active aging in China.

CONCLUSION

The advancement of techniques has driven the openness and sharing of large-scale datasets. Various government data and large-scale social survey data have been opened online worldwide. In China, various data platforms and data sources are opened, for example, China general social survey (CGSS), Chinese Family Panel Studies (CFPS), China Health and Retirement Longitudinal Study (CHARLS), China Education Panel Survey (CEPS), China Household Finance Survey (CHFS), etc. These datasets not only provide important basis for government to make policy and adjust strategic planning, but also offer an important means to advance scientific research. Previously, research on these datasets predominantly use traditional analytic methods such as simple linear regression by first proposing hypotheses and then testing them. In this case, the dataset may not be fully explored and some hidden knowledge may not be revealed. Because machine learning technique can make up for this disadvantage, there has been a rapidly increasing of using machine learning in social science. Machine learning technology is now used in various areas, such as medical assistance diagnosis, loan issuance, and even court trials. By combining complex machine learning models with interpretable machine learning techniques, we can make the best use of big-scale datasets, and gain a deeper understanding of the data.

In the current research, we use the China Longitudinal Health and Longevity Survey (CLHLS) 2014 dataset to investigate factors influencing health status and life satisfaction of elderly in China. We are among the first to use machine learning technique in exploring the predicting factors for health and life satisfaction in CLHLS. We find that positive and optimistic attitudes, a healthy diet, and economic status are common predictors for self-rated health and life satisfaction. Moreover, there are some unique factors influencing self-rated health and life satisfaction. Chronic diseases and mobility abilities greatly affect elderly's self-rated health, and physical examination indicators, living condition, and health-related indicators affect elderly's life satisfaction. We hope that the present research sheds light on the healthy aging and public health literature, and furthermore, provides novel ideas and methods for the use of public large-scale datasets.

REFERENCES

Ali, T., Nilsson, C. J., Weuve, J., Rajan, K. B., & de Leon, C. F. M. (2018). Effects of social network diversity on mortality, cognition and physical function in the elderly: A longitudinal analysis of the Chicago Health and Aging Project (CHAP). *Journal of Epidemiology and Community Health*, *72*(11), 990–996. doi:10.1136/jech-2017-210236 PMID:29970598

Bae, S. Y., Ko, D. S., Noh, J. S., Lee, B. H., Park, H. S., & Park, J. (2010). Relation of physical activity and health-related quality of life in Korean elderly. *The Journal of the Korea Contents Association*, *10*(10), 255–266. doi:10.5392/JKCA.10.10.255

Bähler, C., Huber, C. A., Brüngger, B., & Reich, O. (2015). Multimorbidity, health care utilization and costs in an elderly community-dwelling population: A claims data based observational study. *BMC Health Services Research*, *15*(1), 23. doi:10.118612913-015-0698-2 PMID:25609174

Bind, M. A., Peters, A., Koutrakis, P., Coull, B., Vokonas, P., & Schwartz, J. (2016). Quantile regression analysis of the distributional effects of air pollution on blood pressure, heart rate variability, blood lipids, and biomarkers of inflammation in elderly American men: The Normative Aging Study. *Environmental Health Perspectives*, *124*(8), 1189–1198. doi:10.1289/ehp.1510044 PMID:26967543

Borg, C., Hallberg, I. R., & Blomqvist, K. (2006). Life satisfaction among older people (65+) with reduced self-care capacity: The relationship to social, health and financial aspects. *Journal of Clinical Nursing*, *15*(5), 607–618. doi:10.1111/j.1365-2702.2006.01375.x PMID:16629970

Bose, I., & Mahapatra, R. K. (2001). Business data mining—A machine learning perspective. *Information & Management*, *39*(3), 211–225. doi:10.1016/S0378-7206(01)00091-X

Choi, S. H., Ku, E. J., Hong, E. S., Lim, S., Kim, K. W., Moon, J. H., ... Jang, H. C. (2015). High serum adiponectin concentration and low body mass index are significantly associated with increased all-cause and cardiovascular mortality in an elderly cohort,"adiponectin paradox": The Korean Longitudinal Study on Health and Aging (KLoSHA). *International Journal of Cardiology*, *183*, 91–97. doi:10.1016/j.ijcard.2015.01.057 PMID:25662058

Chou, K. L., & Chi, I. (2001). Stressful life events and depressive symptoms: Social support and sense of control as mediators or moderators? *International Journal of Aging & Human Development*, *52*(2), 155–171. doi:10.2190/9C97-LCA5-EWB7-XK2W PMID:11352200

Chou, K. L., Chi, I., & Chow, N. W. S. (2004). Sources of income and depression in elderly Hong Kong Chinese: Mediating and moderating effects of social support and financial strain. *Aging & Mental Health*, *8*(3), 212–221. doi:10.1080/13607860410001669741 PMID:15203402

De Leon, C. M., Kasl, S. V., & Jacobs, S. (1994). A prospective study of widowhood and changes in symptoms of depression in a community sample of the elderly. *Psychological Medicine*, *24*(3), 613–624. doi:10.1017/S0033291700027768 PMID:7991743

De Martinis, M., Sirufo, M. M., & Ginaldi, L. (2017). Allergy and aging: An old/new emerging health issue. *Aging and Disease*, *8*(2), 162. doi:10.14336/AD.2016.0831 PMID:28400983

Gao, M., Kuang, W., Qiu, P., Wang, H., Lv, X., & Yang, M. (2017). The time trends of cognitive impairment incidence among older Chinese people in the community: Based on the CLHLS cohorts from 1998 to 2014. *Age and Ageing*, *46*(5), 787–793. doi:10.1093/ageing/afx038 PMID:28369164

Giacalone, D., Wendin, K., Kremer, S., Frøst, M. B., Bredie, W. L., Olsson, V., ... Risvik, E. (2016). Health and quality of life in an aging population–Food and beyond. *Food Quality and Preference*, *47*, 166–170. doi:10.1016/j.foodqual.2014.12.002

Goldman, D. P., Shang, B., Bhattacharya, J., Garber, A. M., Hurd, M., Joyce, G. F., ... Shekelle, P. G. (2005). Consequences Of Health Trends And Medical Innovation For The Future Elderly: When demographic trends temper the optimism of biomedical advances, how will tomorrow's elderly fare? *Health Affairs*, *24*(Suppl2), W5–R5. doi:10.1377/hlthaff.W5.R5

Gu, D., Deng, S., Zheng, Q., Liang, C., & Wu, J. (2019). Impacts of case-based health knowledge system in hospital management: The mediating role of group effectiveness. *Information & Management*, *56*(8), 103162. doi:10.1016/j.im.2019.04.005

Hardy, S. E., Concato, J., & Gill, T. M. (2002). Stressful life events among community-living older persons. *Journal of General Internal Medicine*, *17*(11), 841–847. doi:10.1046/j.1525-1497.2002.20105.x PMID:12406354

Hosseini, S. H., & Bahraminejad, Z. (2014). The role of social support networks in public health and health service utilization among the elderly. *J Res Health*, 955-61.

Hu, H. S., & Lu, Y. P. (2012). Income Inequality, Health and Subjective Well-being among the Elderly: Empirical Evidence on the Aging of China. *China Soft Science*, *11*, 41–56.

Huang, W., Lei, X., Ridder, G., Strauss, J., & Zhao, Y. (2013). Health, height, height shrinkage, and SES at older ages: Evidence from China. *American Economic Journal. Applied Economics*, *5*(2), 86–121. doi:10.1257/app.5.2.86 PMID:26594311

Jahangiri, A., & Rakha, H. A. (2015). Applying machine learning techniques to transportation mode recognition using mobile phone sensor data. *IEEE Transactions on Intelligent Transportation Systems*, *16*(5), 2406–2417. doi:10.1109/TITS.2015.2405759

Jiang, H. (2016). Active and healthy aging: The EU's aging social health care strategy and as a reference. *Journal of Southwest University for Nationalities*, *37*(8), 106–111.

Jin, Y., Zhou, F., & Zhai, Z. (2017) The effect of living style on psychological health of the elderly--The moderating effect of community environment. *Population Journal*, (3), 66-77.

Kang, Y., Kim, M., & Lee, E. (2008). The relationship of perceived health status, activities of daily living and nutrition status in the community-dwelling Korean elderly. *Journal of Korean Academy of Nursing*, *38*(1), 122–130. doi:10.4040/jkan.2008.38.1.122 PMID:18323725

Kwon, M. J., & Kim, Y. J. (2016). Analysis of convergent factors on subjective health status of patients with depression. *Journal of Digital Convergence*, *14*(6), 309–316. doi:10.14400/JDC.2016.14.6.309

Li, J., Feng, Y., & Yang, P. (2014). A study on life quality of empty nest elderly in rural areas. *Aging Science Research, 2*(5), 3.

Li, W., Hu, H., & Li, S. (2018). Social activities and health of the elderly: an investigation based on tracking data from 2005 to 2014. *Population and Development*, (2), 90-100.

Liang, C., Gu, D., Tao, F., Jain, H. K., Zhao, Y., & Ding, B. (2017). Influence Mechanism of Patient-Accessible HIS Implementation on Doctor-Patient Relationships: A Service Fairness Perspective. *Information & Management*, *54*(1), 57–72. doi:10.1016/j.im.2016.03.010

Liu, W & Liu, C, (2018) Social insurance and elderly health in the rural area: Can participation in social insurance improve the health of elderly? An empirical study based on multivariate ordered logistic regression. *Social Security Research*, (2), 47-53.

Lu, J, & Wang, X. (2017). Discussion on the bottleneck problem and reform path of restricting the deep integration of medical care. *Population and Birth Control*, (7), 12-14.

Obermeyer, Z., & Emanuel, E. J. (2016). Predicting the future—Big data, machine learning, and clinical medicine. *The New England Journal of Medicine*, *375*(13), 1216–1219. doi:10.1056/NEJMp1606181 PMID:27682033

Perez Martinez, G., Bäuerl, C., & Collado, M. C. (2014). Understanding gut microbiota in elderly's health will enable intervention through probiotics. *Beneficial Microbes*, *5*(3), 235–246. doi:10.3920/BM2013.0079 PMID:24889891

Rolls, E. T., Kellerhals, M. B., & Nichols, T. E. (2015). Age differences in the brain mechanisms of good taste. *NeuroImage*, *113*, 298–309. doi:10.1016/j.neuroimage.2015.03.065 PMID:25842291

Shen, K., & Zeng, Y. (2011). The association between resilience and survival among Chinese elderly. In *Resilience in Aging* (pp. 217–229). New York, NY: Springer. doi:10.1007/978-1-4419-0232-0_14

Shi, Z., Zhang, T., Byles, J., Martin, S., Avery, J., & Taylor, A. (2015). Food habits, lifestyle factors and mortality among oldest old Chinese: The Chinese Longitudinal Healthy Longevity Survey (CLHLS). *Nutrients*, *7*(9), 7562–7579. doi:10.3390/nu7095353 PMID:26371039

Steptoe, A., Wright, C., Kunz-Ebrecht, S. R., & Iliffe, S. (2006). Dispositional optimism and health behaviour in community-dwelling older people: Associations with healthy ageing. *British Journal of Health Psychology*, *11*(1), 71–84. doi:10.1348/135910705X42850 PMID:16480556

Wagnild, G. (2003). Resilience and successful aging: Comparison among low and high income older adults. *Journal of Gerontological Nursing*, *29*(12), 42–49. doi:10.3928/0098-9134-20031201-09 PMID:14692243

Wang, B., Xie, Q., & Yan, H. (2016). Study on the Necessity and Method of Clinical Nursing Path for Elderly Health. *Chinese Digital Medicine*, *11*, 12–14.

Wang, J., Gong, Q., & Wang, W. (2012). Economic Research on Ageing Health. *Economic Research Journal*, *1*, 134–150.

World Health Organization. (2015). World Report on Ageing and Health, WHO Library Cataloguing-in-Publication Data. *World Health Organization*. Retrieved from https://www.who.int/about/who-we-are/constitution

Xue, X, & Ge, K. (2017). The Influence of Social Economic Status on the Health Status of the Elderly-An Empirical study of the Factors Affecting the Elderly Health in China. *Population and Development, 2*, 61-69.

Yan, Y., Zhang, Y., Yang, Q., Dong, K., Duan, F., Liang, S., ... Wang, K. (2018). Association between resting heart rate and hypertension in Chinese with different waist-to-height ratio: A population-based cross-sectional study. *Journal of the American Society of Hypertension*, *12*(12), e93–e101. doi:10.1016/j.jash.2018.08.005 PMID:30268408

Yu, S. J., Kang, C. Y., & Kim, Y. R. (2015). Correlation of quality of life in the health condition of Korean elderly: Perceived, physical, mental health status. *The Korean Journal of Health Service Management*, *9*(2), 47–56. doi:10.12811/kshsm.2015.9.2.047

Zeng, Y., Feng, Q., Gu, D., & Vaupel, J. W. (2017). Demographics, phenotypic health characteristics and genetic analysis of centenarians in China. *Mechanisms of Ageing and Development*, *165*, 86–97. doi:10.1016/j.mad.2016.12.010 PMID:28040447

Zeng, Y., Poston, D. L., Vlosky, D. A., & Gu, D. (Eds.). (2008). *Healthy longevity in China: demographic, socioeconomic, and psychological dimensions* (Vol. 20). Springer Science & Business Media.

Zhiyan. (2017). *Report of in-depth research of the pension industry and investment strategy in China*. Retrieved from https://www.chyxx.com/research/201703/508298.html

Zhong, X. (2013). Eco-economic coupling mechanism in the process of healthy aging and active aging. *China Public Health Management*, *3*, 304–308.

Zhu, Y., Xie, C., Wang, G. J., & Yan, X. G. (2017). Comparison of individual, ensemble and integrated ensemble machine learning methods to predict China's SME credit risk in supply chain finance. *Neural Computing & Applications*, *28*(1), 41–50. doi:10.100700521-016-2304-x

Chapter 3
Urban Solid Waste Management Techniques With Special Reference to Vermicomposting

Junaid Ahmad Malik
https://orcid.org/0000-0003-4411-2015
Government Degree College, Bijbehara, India

ABSTRACT

The majority of municipal (urban) solid waste (MSW) is disposed of in landfills (anaerobic composting). However, this disposal system is reported to produce hazardous environmental impacts, and new policies are initiated to protect the environment from such impacts by discouraging the practice of disposal of solid waste in landfills. Eco-friendly disposal alternatives to landfills need to be explored. One of the technological options for treatment and disposal of organic solid wastes is vermicomposting. Instead of using conventional composting methods, the vermicomposting method can be used for the disposal of the waste. The enzymatic and microbial activities that occur during vermicomposting process by worms gives a better end product than using the conventional composting. Food waste will be recovered for vermicomposting, and the end products can be organic fertilizer, soil improver, and an alternative source of potential energy. Vermicomposting or vermiculture is an eco-friendly process whereby worms are used to breakdown the organic waste into soil and humus known as vermicast, vermicompost, or earthworm compost. The objective is to process the organic material as quickly and efficiently as possible. During the breakdown of organic matter by the earthworm, humus, which is a complicated material, is formed. The material formed is high in content of nitrogen, calcium, and phosphorus, which are the main elements required by plants for their growth, health, and freshness.

DOI: 10.4018/978-1-7998-1966-0.ch003

INTRODUCTION

A fast expanding populace and high rate of industrialization has expanded the issue of solid waste management. The issue has additionally expanded in urban areas on account of lack of dumping destinations and severe natural enactment, so researchers are looking for the executive choices, which ought to be ecofriendly, shoddy and quick. Urban solid waste (MSW) is very carbon-based in nature, so vermicomposting has turned into a suitable option for the innocuous, sterile and cost viable transfer of it. Worms feed on the organics what's more, convert material into casting (expelled material) rich in plant supplements. The compound examinations of these expelled materials show two times accessible magnesium, multiple times available nitrogen and multiple times available potassium contrasted with the encircling soil (Bridgens, 1981).

Urban/municipal solid waste (MSW) is usually regarded as the waste that is generated from human settlements, small industries, commercial and municipal activities (Table 1). There are few more sources from which MSW originates i.e. waste water treatment plants (sewage sludge), household (glass, paper, metals, etc.), public areas (waste from parks, streets, etc.) (Table 1). The management of solid waste has turned out to be one of the most concerning issues we are confronting today. The quick increment in the volume of waste is one viewpoint of the natural emergency, going with later worldwide advancement. Solid waste is considered as the natural and inorganic waste materials created by various sources and have lost an incentive in the eye of their proprietor. It has been evaluated that India creates as much as 25 million tonns of urban solid waste of various formations every year. World urban communities create roughly 1.3 billion metric huge amounts of solid waste every year, which is practically twofold the sums that were produced 10 years back (Hoornweg and Bhada-Tata, 2012). By 2025, strong waste ages will twofold once more (Hoornweg et al., 2013). The yearly increment in solid waste age is inseparably connection to the fast ascent in worldwide populace and rate of urbanization. As a nation urbanizes, its way of life and income level builds which prompts higher utilization of merchandise and enterprises, along these lines creating bigger measure of solid waste per capita (Hoornweg and Bhada-Tata, 2012). Amongst the total solid waste created all inclusive, organic waste is the biggest extent with 46% (Hoornweg and Bhada-Tata, 2012). The organic waste incorporates sustenance scraps, yard waste and agrarian waste. The remainder of the waste is inorganic like paper, plastic, glass, metal and others (Karak et al., 2012). Be that as it may, per capita waste generation in India is minisculous contrasted with the per capita generation of squanders in the industrialized nations. It is assessed that the per capita waste produced in India is about 0.4 kg/day with the compostable issue around 50-60%. In developed nations, the landfills are outfitted with a blend of waste regulation frameworks, for example, spill discovery and the board frameworks for gathering leachates and biogas (Hoornweg and Bhada-Tata, 2012). The impediment of land and the valuation of waste as assets are solid motivations to move away from landfilling and move towards increasingly maintainable waste administration methodology (Marshall and Farahbaksh, 2013). Most basic practices of waste formulating are uncontrolled dumping which causes fundamentally water and soil adulteration. Current techniques for solid waste administrations are landfilling, burning, reusing, reuse, source decrease and others (Wu et al., 2014). Other than dumping or clean land filling, the last transmission of solid waste can be completed by different techniques like burning or incineration and treating the soil by composting. Both landfilling and burning are portrayed as waste transfers, which are the least favoured choices in the waste the board chain of command. In numerous pieces of the world, landfilling remains the predominant technique for waste transfer as it is the least expensive as far as capital expenses (Laner et al., 2012). Farming of Earthworm (vermiculture) is another bio-procedure for changing over the solid organic waste into manure or compost (Ghosh, 2004). Decreased landfilling for expanded reusing of materials (glass, paper, cardboard, plastic, metals and organic chunk) appears to prompt lower ecological effect and lower vitality request or the energy demand (Eriksson et al., 2005).

Vermicomposting is one of the methodologies for mustering the carbon-based portion of community solid wastes for meeting crop nutritive requirements, hence establishes a fecund waste management approach (Sim and Wu, 2010). An inventive discipline of vermiculture biotechnology, reproducing and engendering of earthworms and the utilization of its castings has become a significant device of waste reusing the world over. Basically, the vermiculture offers for the utilization of earthworms as the regular bioreactors for financially savvy and ecologically stable squander management. Various references are accessible on the capability of earthworms in vermicomposting of solid waste especially family squander (Appelhof et al., 1998). Propelled frameworks for vermicomposting depend on top bolstering and base release of a raised reactor, in this way giving solidness and authority over the key zones of temperature, dampness and aerobicity (Price, 1988). An improved mechanical separator, having a novel brushing activity, for expelling live worms from vermicomposts has additionally been presented by Price and Phillips (1990).

The activity of worms during the time spent vermicomposting of waste is physical and biochemical. The physical procedure incorporates substrate air circulation, blending as well as real crushing while the biochemical procedure is impacted by microbial disintegration of substrate in the digestive tract of these worms (Hand et al., 1998). Different studies have demonstrated that vermicomposting of organic waste quickens organic matter stabilisation (Neuhauser et al., 1998; Frederickson et al., 1997) and gives chelating and phytohormonal components (Tomati et al., 1995) which have a high substance of microbial matter and settled humic substances.

Investigating the perilous effect of the inappropriate transfer of squanders on nature, accentuation ought to be given on oxygen consuming fertilizing the soil which changes over waste into natural excrement wealthy in plant supplements and humus (Singh and Sharma, 2002; Sarma et al., 2010; Fernández-Gómez et al., 2010). Transformation of organic squanders into VC has hence picked up significance as of late in light of the fact that of the safe quickened breakdown of natural squanders through microorganisms and worms in mesophilic conditions (Dominguez, 2004).

Vermicompost is known as a sustainable source of macro and micronutrients, plant growth hormones, suppressive microbes and enzymes (Kale and Karmegam, 2010; Sarma et al., 2010) which not only enhance microbial population but also hold nutrients for longer periods (Ndegwa and Thompson, 2001). Vermicompost contain plant-growth regulating materials, such as auxins, gibberellins and cytokinins (Krishnamoorthy and Vajrabhiah, 1986) and humic acids (Atiyeh et al., 2002) which are responsible for increased plant growth and yield of many crops (Atiyeh et al., 2002).

WASTE MANAGEMENT TECHNIQUES

Waste management is the collection, transportation, processing, treatment, recycling or disposal of waste materials to reduce their adverse effects on human health or amenities (www.wastewikipedia.com). Waste can be liquid, gas or a solid substance and it is an unwanted material the owner is about to discard or has discarded. The management of waste in developing countries differs greatly from what is done in advanced communities and also from urban to rural and from residential to industrial settings. The developing country like Nigeria, there are less of metal and plastic wastes compared to high organic wastes (Cointreau, 1982). Political and economic framework is a major problem for the waste management in developing countries. Non-hazardous residential and institutional waste is the responsibility of the local government; even though they claimed responsibility of spending huge amount of money on waste related issues. The type of waste management techniques that should be applied for proper management of waste depend on the composition of the waste. Although composting will be appropriate for all organic wastes.

1. Landfilling

It is an economical method of waste disposal in developing countries involving pitching refuses into a depression, abandoned, mining void, excavated land or borrowed pits (Daskalopoulos et al. 1998). Landfilling is the most traditional way of true waste disposal practised in many countries. The process involves the reclamation of land in low-lying areas by addition of gravel, rubble, or other loose material. Various forms of landfills are;

a. ***Open dump System or Ordinary landfill:*** this disposal of waste materials is in pits, canals and sloping landscapes without covering the waste. An open dumping is defined as a land disposal site at which solid wastes are disposed of in a manner that does not protect the environment, are susceptible to open burning, and are exposed to the elements, vectors, and scavengers. Open dumping can include solid waste disposal facilities or practices that pose a reasonable probability of adverse effects on health or the environment. Abandoned piles of household garbage, bags of yard waste, appliances, old barrels, used tires and demolition debris such as lumber, shingles, pipes and asbestos can threaten the health of humans, wildlife and the environment. Known as open dumps, these sites can be found throughout Illinois – heaped at the bottom of ravines, in empty lots and pastures and along roadsides. An open dump is an illegal waste disposal site and should not be confused with a permitted municipal solid waste landfill or a recycling facility. If allowed to remain, open dumps often grow large and may attract the dumping of both solid and hazardous wastes.
b. ***Sanitary Landfill:*** This engineering means of disposing solid wastes uses thin layers, compacted into the smallest practical volume and cover with inert ash at the end of each working day. Sanitary landfills are sites where waste is isolated from the environment until it is safe. It is considered when it has completely degraded biologically, chemically and physically. In high-income countries, the level of isolation achieved may be high. However, such an expensive high level of isolation may not be technically necessary to protect public health. Four basic conditions should be met before a site can be regarded as a sanitary landfill. The ways of doing this should be adapted to local conditions. The immediate goal is to meet, to the best extent possible, the four stated basic sanitary landfill conditions, with a longer term goal to meet them eventually in full. As a minimum, four basic conditions should be met by any site design and operation before it can be regarded as a sanitary landfill:
    i. Full or partial hydrogeological isolation: if a site cannot be located on land which naturally contains leachate security, additional lining materials should be brought to the site to reduce leakage from the base of the site (leachate) and help reduce contamination of groundwater and surrounding soil. If a liner - soil or synthetic - is provided without a system of leachate collection, all leachate will eventually reach the surrounding environment. Leachate collection and treatment must be stressed as a basic requirement.

ii. Formal engineering preparations: designs should be developed from local geological and hydrogeological investigations. A waste disposal plan and a final restoration plan should also be developed.
iii. Permanent control: trained staff should be based at the landfill to supervise site preparation and construction, the depositing of waste and the regular operation and maintenance.
iv. Planned waste emplacement and covering: waste should be spread in layers and compacted. A small working area which is covered daily helps make the waste less accessible to pests and vermin.

c. ***Secured Landfill:*** A secure landfill is a carefully engineered area that is used to deposit waste products. Typically, a secure landfill is a hole in the ground, but may also be built above ground. If the depression is in the ground, it must provide a 3 meter (10 foot) separation between the bottom of the landfill and the underlying bedrock or groundwater table. This landfill is mainly constructed for hazardous wastes disposal such as hospital and radioactive wastes. It is not designed properly, the resultant effect could be similar to that of a sanitary landfill.

2. Incineration

Incineration is a waste treatment process that involves the combustion of organic substances contained in waste materials (Knox and Andrew, 2005). Incineration and other high-temperature waste treatment systems are described as "thermal treatment". Incineration of waste materials converts the waste into ash, flue gas and heat. The ash is mostly formed by the inorganic constituents of the waste and may take the form of solid lumps or particulates carried by the flue gas. The flue gases must be cleaned of gaseous and particulate pollutants before they are dispersed into the atmosphere. In some cases, the heat generated by incineration can be used to generate electric power. Incineration with energy recovery is one of several waste-to-energy technologies such as gasification, pyrolysis and anaerobic digestion. While incineration and gasification technologies are similar in principle, the energy produced from incineration is high-temperature heat whereas combustible gas is often the main energy product from gasification. Incineration and gasification may also be implemented without energy and materials recovery. Incinerators reduce the solid mass of the original waste by 80–85% and the volume (already compressed somewhat in garbage trucks) by 95–96%, depending on composition and degree of recovery of materials such as metals from the ash for recycling (Herbert and Lewis, 2007). An incinerator is a furnace for burning waste. Modern incinerators include pollution mitigation equipment such as flue gas cleaning. There are various types of incinerator plant design: moving grate, fixed grate, rotary-kiln, and fluidised bed. Incineration has particularly strong benefits for the treatment of certain waste types in niche areas such as clinical wastes and certain hazardous wastes where pathogens and toxins can be destroyed by high temperatures. Examples include chemical multi-product plants with diverse toxic or very toxic wastewater streams, which cannot be routed to a conventional wastewater treatment plant. The benefits of incineration are a major reduction in waste volume and production of energy and heat production (Seo et al., 2004).

3. Pyrolysis and Gasification

Combustion by means of incineration can be one of the solutions for recovering the energy content of waste, however it involves numerous difficulties, among which the emission of gas effluents that require a costly purification treatment and that have induced researchers and engineers to search for more solutions for the plants. Among these are gasification and pyrolysis, which are being experimented as a potential alternative to the Waste to Energy systems. Even if in the waste sector, innovative technologies are being considered, gasification and pyrolysis have a more ancient history that dates back to the 18th century. These are methods for managing wastes by heating under controlled conditions to produce low to medium heating fuel gases, tars, char and ash; under a high temperature with limited oxygen (Heimlich et al., 2005). Usually the process takes place in a sealed vessel under a high pressure. Whereas pyrolysis converts the solid wastes into solid, liquid and gas products, gasification converts organic materials into a syngas (CO and H_2).

4. Land Application of Waste

This process involves the direct land application of organic waste in agricultural area without any pretreatment. Land application of organic waste materials such as such as sewage sludge provides valuable nutrients and helps in organically enriching soils and restoring the opportunity for improved plant growth (Singh and Agrawal, 2009, 2010a). Organic waste from municipal area usually contains high organic matter, micronutrients, nitrogen, phosphorous, potassium (Sigua, 2005; González et al., 2008). The macronutrients present in waste serves as a good source of plant nutrients and the organic constituents provide beneficial soil conditioning properties (Singh and Agrawal, 2010b). Heavy metals are also abundant in waste due to mixing of industrial wastes and changing life style (McGrath et al., 2000). This is cause of main concern. As their long-term use can cause heavy metal accumulation in soil (Lopez-Mosquera et al., 2000). Once accumulated in soil, heavy metals may be transferred at elevated levels to the food chain (Page et al., 1987), which may pose a variety of human health problems (Wang et al., 2003).

THE WORMS

Nature has produced and segregated earthworms into epigeic, anecic and endogeic species based on definite ecological and trophic meanings in the respective soils. Of the three ecological varieties, the epigeics in particular and the aneceic in general, have largely been harnessed for use in the vermicomposting process. Epigeics like *Eisenia foetida* (Hartenstein et.al., 1979) and *Eudrilus euginiae* (Kale and Bano, 1988) have been used in converting organic wastes (agro waste and domestic refuse) into vermicompost. Though these surface dwellers are capable of working hard on the litter layer and can convert all the organic waste into manure they are of no significant value in modifying the structure of the soil. The anecics however are capable of both organic waste consumption as well as in modifying the structure of the soil. Such burrowing species that are widely used in soil management like the earthworms, *Lampito mauritii* (Ismail, 1993) also effectively create a drilosphere apart from helping in compost production. Worm-worked soils are conspicuously different from soil devoid of earthworms. Importantly, the tunnel formed by the worms (Rahuman, 1984) aid in the passage of water, which also washes the nutrients from the drilosphere to the roots that extend quite rapidly along these channels. This principle is also applied in the preparation of vermiwash. Vermiwash functions as a good foliar spray, as it contains the requisite nutrients in it (Ismail, 1997).

NECESSARY TRAITS OF VERMICOMPOSTER WORMS

1. Worm should exhibit high biomass consumption together with a high efficiency of conversion of ingested biomass to body proteins, a physiological trait required for achieving high growth rate.
2. Worm should have wider range of tolerance to environmental factors including adaptation to feed on a variety of organic residues
3. Worm should produce large numbers of cocoons with short hatching time enabling rapid population growth and, linked to this rapid growth, faster composting of organic residues.
4. Life cycle of the worm should be such that mature/adult phase is quickly reached.
5. Using a mixture of species is likely to be more useful than use of single species.
6. Worm should be disease resistant.

The Difference Between Vermiculture and Vermicomposting

Vermiculture is the culture of worms. The objective is to persistently expand the quantity of worms so as to acquire a manageable collect. The worms are either used to extend a vermicomposting activity or offered to clients who use them for the equivalent or different purposes.

Vermicomposting is the procedure by which worms are utilized to change over natural materials (normally squanders or wastes) into a humus-like material known as vermicompost. The objective is to process the material as fast and effectively as could reasonably be expected.

These two procedures are comparative however extraordinary. On the off chance that you will probably convey vermicompost, you will need to have your most extreme worm populace thickness constantly. On the off chance that you will probably create worms, you will need to keep the populace thickness low enough that regenerative rates are upgraded.

VERMICOMPOSTING PROCESS

It is an aerobic, bio-oxidation, non-thermophilic process of organic waste decomposition that depends upon earthworms to fragment, mix and promote microbial activity.). It has been estimated that earthworms add 230 kg N/ha/year in grasslands and 165 kg N/ha/year in woodland sites. Earthworms increase the nitrate production by stimulating bacterial activity and through their own decomposition. There are reports that concentrations of exchangeable cations such as Ca, Mg, Na, K, available P and Mo in the worm casts are higher than those in the surrounding soil. Vermicompost cannot be described as being nutritionally superior to other organic manures. Instead, it is a unique way of manure production. The basic requirements during the process of vermicomposting are;

- Suitable bedding
- Food source
- Adequate moisture
- Adequate aeration
- Suitable temperature
- Suitable Ph

Bedding

Bedding is any material that provides a relatively stable habitat to worms. If available, shredded paper or cardboard makes excellent bedding, particularly when combined with typical on-farm organic resources such as straw and hay. For good vermicomposting, this habitat should satisfy the following criteria;

High Absorbency: As worms breathe through skin, the bedding must be able to absorb and retain adequate water.

Good Bulking Potential: The bulking potential of the material should be such that worms get oxygen properly.

Low Nitrogen Content (High Carbon: Nitrogen Ratio): Although worms consume their bedding as it breaks down, it is very important that this be a slow process. High protein/nitrogen levels can result in rapid degradation and associated heating may be fatal to worms.

Food Source

Regular input of feed materials for the earthworms is most essential step in the vermicomposting process. Earthworms can use a wide variety of organic materials as food but do exhibit food preferences. In adverse conditions, earthworms can extract sufficient nourishment from soil to survive. However earthworms feed mainly on dead and decaying organic waste and on free living soil microflora and fauna. Under ideal conditions, worms can consume amount of food higher than their body weights, the general rule-of-thumb is consumption of food weighing half of their body weight per day. Livestock excreta, viz., goat manure, cattle dung or pig manure are the most commonly used worm feedstock as these materials have higher nitrogen content. When the material with higher carbon content is used with C:N ratio exceeding 40:1, it is advisable to add nitrogen supplements to ensure effective decomposition. The food should be added only as a limited layer as an excess of the waste many generate heat. From the waste ingested by the worms, 5-10% are being assimilated in their body and the rest are being excreted in the form of vermicast.

Moisture

Perhaps the most important requirement of earthworms is adequate moisture. They require moisture in the range of 60-70%. The feed stock should not be too wet otherwise it may create anaerobic conditions which may be fatal to earthworms.

Aeration

Factors such as high levels of fatty/oily substances in the feedstock or excessive moisture combined with poor aeration may render anaerobic conditions in vermicomposting system. Worms suffer severe mortality partly because they are deprived of oxygen and partly because of toxic substances (e.g. ammonia) produced under such conditions. This is one of the main reasons for not including meat or other fatty/oily wastes in worm feedstock unless they have been pre-composted to break down the oils and fats.

Temperature

The activity, metabolism, growth, respiration and reproduction of earthworms are greatly influenced by temperature. Most earthworm species used in vermicomposting require moderate temperatures from 10-35°C. While tolerances and preferences vary from species to species. Earthworms can tolerate cold and moist conditions far better than hot and dry conditions. For *Eisenia foetida* temperatures above 10°C (minimum) and preferably 15°C be maintained for maximizing vermicomposting efficiency and above 15°C (minimum) and preferably 20°C for vermiculture. Higher temperatures (>35°C) may result in high mortality. Worms will redistribute themselves within piles, beds or windrows such that they get favorable temperatures in the bed.

pH

Worms can survive in a pH range of 5 to 9, but a range of 7.5 to 8.0 is considered to be the optimum. In general, the pH of worm beds tends to drop over time due to the fragmentation of organic matter under series of chemical reactions. Thus, if the food sources are alkaline, the effect is a moderating one, tending to neutral or slightly acidic, and if acidic (e.g., coffee grounds, peat moss); pH of the beds can drop well below 7. In such acidic conditions, pests like mites may become abundant. The pH can be adjusted upwards by adding calcium carbonate.

COMPOSTING AND VERMICOMPOSTING FORMS

Composting is a biological decomposition of organic waste either in an aerobic (Makan et al., 2014) or anaerobic (Minale and Worku, 2014) environment with the former being more common. Several studies have been reported from around the world as well in Iran, in terms of consumption of organic residue such as sewage sludge, animal remains, industrial and agricultural wastes by earthworms and converting it to 'vermicompost' on a small-scale methods. The organic matters in the waste are consumed by aerobic thermophilic and mesophilic microorganisms as substrates and converted into mineralized products such as CO_2, H_2O, NH_4^+ or stabilized organic matters (Qian et al., 2014). The resultant compost is a stable, humus-rich, complex mixture that can improve physical properties of the soil (Watteau and Villemin, 2011). Factors affecting composting process are temperature, initial C/N ratio, aeration, porosity, moisture content and pH (Shafawati and Siddiquee, 2013). During composting process, these parameters are regulated and controlled to provide an optimum environment for the microorganisms to degrade the organic waste (López-González et al., 2015).

Similar to composting, vermicomposting process is also a biological decomposition of organic waste to produce stabilized organic fertilizer, namely vermicompost. Unlike composting, vermicomposting process involves interactions between earthworms and microorganisms to biodegrade organic waste at a faster rate (Sim and Wu, 2010). Earthworms act as the main drivers in the decomposition of organic waste by fragmenting and conditioning the substrate. In doing so, earthworms increase the surface area of the organic waste that is exposed to the microorganisms. Thus, the microbial activity and decomposition process of solid waste are enhanced. Vermicomposting results in the production of vermicompost or earthworm cast that has low C/N ratio, high porosity, water-holding capacity and available nutrients (Lim et al., 2015). Like composting, efficiency of vermicomposting process is also influenced by several factors such as initial C/N ratio, moisture content, pH and nature of the organic waste. In comparison with the composting process, all the factors influencing vermicomposting process are also inextricably linked to the earthworm species which are used during the biodegradation process. In addition to the vermicompost, earthworm biomass is also produced during vermicomposting. After the completion of vermicomposting process, earthworms were removed from the vermicompost via light, vertical or sideways separation. The excess earthworms after the vermicomposting process ends could be used to biotransform other organic waste or as a protein source for animals and as fishing bait (Edwards et al., 2010). Earthworms used in the vermicomposting process must possess the following characteristics: 1) high rates of organic matter consumption, digestion and assimilation; 2) high tolerance of environmental stress; 3) high reproductive rate; 4) rapid growth and maturation rate of hatchlings (Singh et al., 2011). Earthworms are classified into three different categories, namely epigeic, endogeic and anecic. Among these earthworm species, epigeic earthworms are the most suitable earthworms to be used in vermicomposting process as they live in organic horizons and feed primarily on decaying organic matter. On the other hand, endogeic earthworms feed on subsurface soil and live below the soil surface while anecic earthworms prefers to feed on soil and lives deep in the soil (Yadav and Garg, 2011). Epigeic species are the most efficient in biodegrading organic waste and releasing nutrients into the soil. In addition, the latest study revealed that surface-dwelling mode of life guarded the epigeic earthworm against their exposure to pesticide (Suthar, 2014). Among the epigeic earthworms, *Eisenia fetida* and *E. andrei* are the most commonly used in vermicomposting because both earthworms are peregrine and ubiquitous with a worldwide distribution, resilient and have wide temperature tolerance (Edwards, 2004).

Composting of Organic Waste

Briefly, composting process can be divided into three phases, namely initial activation, thermophilic and mesophilic or maturation phase (Chowdhury et al., 2013). The majority of the organic waste degradation happens during the thermophilic phase. In this phase, microorganisms degrade the readily available compounds in the organic waste. Generally, high microbial activity translates to high degradation of organic waste (Fourti, 2013). This phase is characterized by high temperature in the composting pile due to the heat released from the microbial catabolism of organic waste (Singh and Kalamdhad, 2014). The high temperature achieved in this phase is also crucial for pathogen reduction and sanitization of organic waste. Temperatures above 55°C are required to kill the pathogens in the organic waste (Tian et al., 2012). The EPA guidelines require composting material to maintain a temperature of 55°C for at least 15 days or 5 consecutive days (Jurado et al., 2014). Some composting piles are known to reach temperatures as high as 70°C during the degradation of animal manure (Tang et al., 2011) and green waste (Cắceres et al., 2015). The end of the thermophilic phase and beginning of the maturation phase is indicated by the temperature decrease in the composting pile. As the organic waste stabilizes, the temperature will continue to decrease to ambient air temperature (Sắnchez-Monedero et al., 2010). The temperature decrease also marks the exhaustion of decomposable organic fraction in the waste (Ravindran and Sekaran, 2010). Some composting studies that were reported did not achieve the thermophilic temperature. For example, Paradelo et al. (2013) noted that the temperatures were in the mesophilic range during composting of lignocellulosic winery waste, but an evidence of organic waste decomposition was found. Low temperature during composting process could be caused by low ambient temperature or heat retention properties of the composting materials (Singh and Kalamdhad, 2014). In fact, low temperature is typical for home composting because the layer for decomposing material is too thin to retain a significant amount of heat and the heat is quickly transferred out from the pile. Although organic waste could be decomposed in low temperature conditions, the end product is not guaranteed to be free of pathogens or weed seeds (Faverial and Sierra, 2014).

Vermicomposting of Organic Waste

Vermicomposting has been attempted on many types of organic solid waste such as animal manure, sewage sludge, vegetable waste, industrial sludge and agriculture waste. Vermicomposting, similar to composting process, is also an organic waste decomposition process but with an addition of earthworms to aid and hasten the waste stabilization process (Lim et al., 2015b). Therefore, suitable organic waste or feedstock for earthworms is crucial to ensure a successful and efficient vermicomposting process (Yadav and Garg, 2011).

Earthworms can consume most organic materials that have pH in the range from 5 to 8, moisture content between 40-55% and initial C/N ratio around 30. However, not all organic waste fall within these parameters. Therefore, to make the organic waste more suitable for vermicomposting, the waste should be: i) amended with bulking agents/organic waste (or amendments) or ii) undergone some form of pre-treatment process. During vermicomposting process, bulking agents or amendments are used to make the organic waste more palatable for the earthworms. For example, cow dung is commonly used as an amendment in vermicomposting process because it is the easiest animal waste for growing the earthworms (Edwards, 2004). Fruit waste could also be used in some instances as amendments during vermicomposting of soybean husk (Lim et al., 2011) and rice husk (Lim et al., 2012). Some organic waste that have high moisture content are amended with bulking agents (or amendments) to reduce the moisture content. In fact, liquid waste like palm oil mill effluent was absorbed onto a bulking material for vermicomposting process (Lim et al., 2014). Other amendments used are soil and matured vermicompost (Huang et al., 2014). Vermicompost is used to provide an initial habitat for earthworms and as a source of microbial inoculums (Castillo et al., 2013). One of the major concerns using vermicomposting process is that it does not involve a thermophilic stage. However, limited studies showed that vermicomposting could reduce pathogens in the organic waste. Yadav et al. (2010) detected no coliforms in the mature vermicompost derived from source-separated human fecal. It was suggested that pathogens were killed or reduced through the action of intestinal enzymes in the earthworms as well as the competition between pathogens and microbes for the limited resources that earthworms left behind (Sim and Wu, 2010). This result was supported by Hill and Baldwin (2012) who proved the viability of using vermicomposting to convert fecal matter and toilet paper into vermicompost which had lesser *Escherichia coli*.

Integration of Composting-Vermicomposting Process

Organic waste stabilization through composting or vermicomposting is well established and proven to be feasible. However, each of this process has its own disadvantages. In the composting process, longer duration is needed for waste stabilization and the organic waste requires frequent aeration to maintain aerobic conditions (Hait and Tare, 2011b). The major drawback during vermicomposting process is the low temperature required for the duration of the process due to the presence of earthworms. The waste might not be sanitized at low temperature and the vermicompost could not meet the required level of pathogen in the organic fertilizer (Fornes et al., 2012). Thus, an integrated composting vermicomposting process is suggested to overcome the disadvantages of each stabilization process. The composting stage in the integrated system will ensure that the produced fertilizer is free from pathogens due to the presence of thermophilic stage, while the vermicomposting stage reduces the particle size and increases available nutrient recovery at a much higher ratem with the presence of earthworms. In addition, earthworms can help to turn and mix the organic waste which will help reduce the cost needed to maintain aerobic conditions during organic matter stabilization (Dominguez and Edwards, 2011).

Besides, the integrated system allows for a wider range of organic waste to be treated. For example, high moisture content of faecal slurry was not suitable to be used as a direct feedstock for vermicomposting process. Therefore, composting stage could be introduced before the pre-treated waste was subjected to vermicomposting process (Yadav et al., 2012). A comparison between composting process alone and the integrated system revealed higher organic matter stability and nutrients (P, Mg and Ca) in the latter (Sierra et al., 2013). Maňáková et al. (2014) showed that higher reduction of arsenic mobility and availability in the integrated system as compared to composting or vermicomposting process alone. Similarly, Soobhany et al. (2015) showed lower Cd, Cr, Cu, Co, Zn and Ni content in the integrated system as compared to composting process alone, suggesting that earthworms reduced the heavy metals in the organic fertilizer by accumulating the metals in their bodies during vermicomposting process. According to Wang et al. (2014), lower greenhouse gas emissions were produced in the integrated system. Therefore, it would seem that the integration of composting and vermicomposting process results in a higher quality organic fertilizer and poses lesser negative impact on the environment. Furthermore, the thermophilic composting stage in the integrated system also ensures the fertilizer fulfilling the pathogen reduction requirements.

STEPS IN THE PROCESS

Vermicomposting involves the following steps;

- Cover the bottom of the cement ring with a layer of tiles or coconut husk or polythene sheet.
- Spread 15–20 cm layer of organic waste material on the polythene sheet. Sprinkle rock phosphate powder if available (it helps in improving nutritional quality of compost) on the waste material and then sprinkle cow dung slurry. Fill the ring completely in layers as described. Paste the top of the ring with soil or cow dung. Allow the material to decompose for 15 to 20 days.
- When the heat evolved during the decomposition of the materials has subsided (15–20 days after heaping), release selected earthworms (500 to 700) through the cracks developed.
- Cover the ring with wire mesh or gunny bag to prevent birds from picking the earthworms. Sprinkle water every three days to maintain adequate moisture and body temperature of the earthworms.
- The vermicompost is ready in about 2 months if agricultural waste is used and about 4 weeks if sericulture waste is used as substrate.
- The processed vermicompost is black, light in weight and free from bad odour.
- When the compost is ready, do not water for 2–3 days to make compost easy for sifting. Pile the compost in small heaps and leave under ambient conditions for a couple of hours when all the worms move down the heap in the bed (Fig. 4i). Separate upper portion of the manure and sieve the lower portion to separate the earthworms from the manure. The culture in the bed contains different stages of the earthworm's life cycle, namely, cocoons, juveniles and adults. Transfer this culture to fresh half decomposed feed material. The excess as well as big earthworms can be used for feeding fish or poultry. Pack the compost in bags and store the bags in a cool place.
- Prepare another pile about 20 days before removing the compost and repeat the process by following the same procedure as described above.

BENEFITS OF VERMICOMPOSTING PROCESS

Like the conventional compost, vermicompost is advantageous to agricultural soil due to increased moisture retention ability, better nutrient holding capacity, superior soil structure, and higher levels of microbial activity. According to Suthar (2009) vermicomposting results in a better quality product in terms of nutrient availability than traditional composting system. Vermiculture technology has a number of benefits, as it is odourless, cost efficient, free of toxic waste and its resultant is a valuable end product. According to Ghosh et al. (1999) vermicomposting might be an efficient technology for providing better P nutrition from different organic wastes. Vermicomposting is an environmentally sustainable process as it leads in destruction of pathogens, small greenhouse emissions, and is scalable to suit any volume. According to Eastman (1999) the process of vermicomposting can also result in a product with a lower pathogen level than that in compost. The nutrient content of the vermicompost depends upon the quality of feed material to earthworm. Earthworm ingests large amount of substrates and are therefore exposed to heavy metals through their skin and intestine. They concentrate heavy metals in their body from the substrates (Morgan and Morgan, 1999).

CONCLUSION

The production of degradable organic waste and its safe disposal becomes the current global problem. Meanwhile the rejuvenation of degraded soils by protecting topsoil and sustainability of productive soils is a major concern at the international level. Provision of a sustainable environment in the soil by amending with good quality organic soil additives enhances the water holding capacity and nutrient supplying capacity of soil and also the development of resistance in plants to pests and diseases. By reducing the time of humification process and by evolving the methods to minimize the loss of nutrients during the course of decomposition, the fantasy

Table 1. General sources of municipal solid waste

Source	Activities, typical amenities, or locations where wastes are generated	Types of solid waste
Residential	Single-family and multi-family home, low, medium, and high rise apartments, etc.	Food wastes, rubbish, paper waste, ashes, special wastes
Commercial and institutional	Warehouses, restaurants, markets, office buildings, hotels, shopping malls, schools, print shops, auto repair shops, medical facilities and institutions, prisons	Food wastes, rubbish, ashes, demolition and construction wastes, special wastes, occasionally hazardous wastes
Open areas	Streets, alleys, parks, vacant lots, playgrounds, beaches, highways, recreational areas, marriage halls, etc.	Street sweepings, roadside litter, rubbish, and other special wastes
Treatment plant sites	Water, sewage and industrial waste water treatment processes	Treatment plant sludges

Source: Peavy et al. (1987).

becomes fact. Earthworms can serve as tools to facilitate these functions. They serve as "nature's plowman" and form nature's gift to produce good humus, which is the most precious material to fulfill the nutritional needs of crops. The utilization of vermicompost results in several benefits to farmers, industries, environment and overall national economy

As a processing system, the vermicomposting of organic waste is very simple. Worms ingest the waste material - break it up in their rudimentary gizzards - consume the digestible/putrefiable portion, and then excrete a stable, humus-like material that can be immediately marketed and has a variety of documented benefits to the consumer. Vermitechnology can be a promising technique that has shown its potential in certain challenging areas like augmentation of food production, waste recycling, management of solid wastes etc. There is no doubt that in India, where on side pollution is increasing due to accumulation of organic wastes and on the other side there is shortage of organic manure, which could increase the fertility and productivity of the land and produce nutritive and safe food. So the scope for vermicomposting is enormous.

REFERENCES

Appelholf, M., Edward's, C.A. & Neuhauser, E.F. (1998). Domestic vermicomposting systems. Earthworm. *Waste Environ. Manage.*, 157-161.

Atiyeh, R. M., Lee, S., Edwards, C. A., Arancon, N. Q., & Metzger, J. D. (2002). The influence of humic acids derived from organic wastes on plant growth. *Bioresource Technology*, *84*, 7–14. doi:10.1016/S0960-8524(02)00017-2 PMID:12137272

Bridgens, S. (1981). The importance of the earthworms. *Span*, *22*(8), 20.

Cáceres, R., Coromina, N., Malińska, K., & Marfà, O. (2015). Evolution of process control parameters during extended co-composting of green waste and solid fraction of cattle slurry to obtain growing media. *Bioresource Technology*, *179*, 396–406. doi:10.1016/j.biortech.2014.12.051 PMID:25553571

Castillo, J. M., Romero, E., & Nogales, R. (2013). Dynamics of microbial communities related to biochemical parameters during vermicomposting and maturation of agroindustrial lignocellulose wastes. *Bioresource Technology*, *146*, 345–354. doi:10.1016/j.biortech.2013.07.093 PMID:23948272

Chowdhury, A. K. M. M. B., Akratos, C. S., Vayenas, D. V., & Pavlou, S. (2013). Olive mill waste composting: A review. *Int. Biodeteriorat. Biodegrad.*, *85*, 108–119. doi:10.1016/j.ibiod.2013.06.019

Cointreau, S. J. (1982). *Environmental Management of Urban solid wastes in developing countries: A Project Guide* (1st ed.). Washington, DC: Wold Bank.

Daskalopoulos, E., Badr, O., & Probert, S. D. (1998). An integrated approach to municipal solid waste management. *Resources, Conservation and Recycling*, *24*(1), 33–50. doi:10.1016/S0921-3449(98)00031-7

Dominguez, J. (2004). State-of-the-art and new perspectives on vermicomposting research. In C. A. Edwards (Ed.), *Earthworm Ecology* (2nd ed.; pp. 401–424). Boca Raton, FL: CRC Press LLC. doi:10.1201/9781420039719.ch20

Dominguez, J., & Edwards, C. A. (2011). Relationships between composting and vermicomposting. In C. A. Edwards, N. Q. Arancon, & R. Sherman (Eds.), *Vermiculture technology* (pp. 11–25). Boca Raton, FL: CRC Press.

Eastman, B. R. (1999). Achieving pathogen stabilization using vermicomposting. *BioCycle*, 62–64.

Edwards, C. A. (2004). *Earthworm Ecology* (2nd ed.). Boca Raton, FL: CRC Press LLC. doi:10.1201/9781420039719

Edwards, C. A., Arancon, N. Q., & Sherman, R. (2010). *Vermiculture Technology: Earthworms, Organic wastes, and Environmental Management*. Boca Raton, FL: CRC Press LLC. doi:10.1201/b10453

Eriksson, O., Carlsson Reich, M., Frostell, B., Björklund, A., Assefa, G., Sundqvist, J. O., ... Thyselius, L. (2005). Municipal solid waste management from a systems perspective. *Journal of Cleaner Production*, *13*(3), 241–252. doi:10.1016/j.jclepro.2004.02.018

Faverial, J., & Sierra, J. (2014). Home composting of household biodegradable wastes under tropical conditions of Guadeloupe (Frence Antilles). *Journal of Cleaner Production*, *83*, 238–244. doi:10.1016/j.jclepro.2014.07.068

Fernández-Gómez, M. J., Nogales, R., Heribert, I., Esperanza, R., & Marta, G. (2010). Continuous-feeding vermicomposting as a recycling management method to revalue tomato-fruit wastes from greenhouse crops. *Waste Management (New York, N.Y.)*, *30*(12), 2461–2468. doi:10.1016/j.wasman.2010.07.005 PMID:20675115

Fornes, F., Mendoza-Hernández, D., García-de-la-Fuente, R., Abad, M., & Belda, R. M. (2012). Composting versus vermicomposting: A comparative study of organic matter evolution through straight and combined processes. *Bioresource Technology, 118*, 296–305. doi:10.1016/j.biortech.2012.05.028 PMID:22705537

Fourti, O. (2013). The maturity tests during composting of municipal solid wastes. *Resources, Conservation and Recycling, 72*, 43–49. doi:10.1016/j.resconrec.2012.12.001

Frederickson, J., Butt, K. R., Morris, R. M., & Danial, C. (1997). Combining vermiculture with traditional green waste composting systems. *Soil Biology & Biochemistry, 29*(3-4), 725–730. doi:10.1016/S0038-0717(96)00025-9

Ghosh, C. (2004). Integrated Vermi-Pisciculture – an alternative option for recycling of solid municipal waste in rural India. *Bioresource Technology, 93*(1), 71–75. doi:10.1016/j.biortech.2003.09.014 PMID:14987723

Ghosh, M., Chattopadhyay, G. N., & Baral, K. (1999). Transformation of phosphorus during vermicomposting. *Bioresource Technology, 69*(2), 149–154. doi:10.1016/S0960-8524(99)80001-7

González, M., Mingorance, M., Sánchez, L., & Pe˜na, A. (2008). Pesticide adsorption on a calcareous soil modified with sewage sludge and quaternary alkyl-ammonium cationic surfactants. *Environmental Science and Pollution Research International, 15*(1), 8–14. doi:10.1065/espr2007.02.387 PMID:18306881

Hait, S., & Tare, V. (2011b). Vermistabilization of primary sewage sludge. *Bioresource Technology, 102*(3), 2812–2820. doi:10.1016/j.biortech.2010.10.031 PMID:21036608

Hand, P., Hayes, W.A., Satchell, J.E., Frankland, J.C., Edwards, C.A. & Neuhauser, E.F. (1998). The vermicomposting of cow slurry. Earthworm. *Waste Environ. Manage.*, 49-63.

Hartenstein, R., Neuhauser, E. F., & Kaplan, D. L. (1979). A progress report on the potential use of earthworms in sludge management. In *Proceedings of the Eighth National Sludge Conference, Florida*. Information Transfer Inc.

Heimlich, J. E., Hughes, K. L., & Christy, A. D. (2005). *Integrated Waste Management*. OSU Extension.

Herbert & Lewis. (2007). *Centenary History of Waste and Waste Managers in London and South East England*. Chartered Institution of Wastes Management.

Hill, G. B., & Baldwin, S. A. (2012). Vermicomposting toilets, an alternative to latrine style microbial composting toilets, prove far superior in mass reduction, pathogen destruction, compost quality, and operational cost. *Waste Management (New York, N.Y.)*, *32*(10), 1811–1820. doi:10.1016/j.wasman.2012.04.023 PMID:22658870

Hoornweg, D., & Bhada-Tata, P. (2012). *What a waste: A Global Review of Solid Waste Management*. Washington, DC: World Bank.

Hoornweg, D., Bhada-Tata, P., & Kennedy, C. (2013). Waste production must peak this century. *Nature*, *502*(7473), 615–617. doi:10.1038/502615a PMID:24180015

Huang, K., Li, F., Wei, Y., Fu, X., & Chen, X. (2014). Effects of earthworms on physiochemical properties and microbial profiles during vermicomposting of fresh fruit and vegetable wastes. *Bioresource Technology*, *170*, 45–52. doi:10.1016/j.biortech.2014.07.058 PMID:25118152

Ismail, S. A. (1993). *Keynote papers and Extended Abstracts. Congress on traditional Sciences and technologies of India*. IIT Bombay.

Ismail, S. A. (1997). *Vermicology: The biology of Earthworms*. Hyderabad: Orient Longman.

Jurado, M. M., Suárez-Estrella, F., Vargas-García, M. C., López, M. J., López-González, J. A., & Moreno, J. (2014a). Evolution of enzymatic activities and carbon fractions throughout composting of plant waste. *Journal o Environmental Management*, *133*, 355–364. doi:10.1016/j.jenvman.2013.12.020 PMID:24412984

Kale, R. D., & Bano, K. (1988). Earthworm cultivation and culturing techniques for production of Vee COMP83E UAS, Vee MEAL 83P UAS. *Mysore Journal of Agricultural Sciences*, *22*, 339–344.

Kale, R. D., & Karmegam, N. (2010). The role of earthworms in tropics with emphasis on Indian ecosystems. *Applied and Environmental Soil Science*, *56*, 41–43.

Karak, T., Bhagat, R. M., & Bhattacharyya, P. (2012). Municipal solid waste generation, composition, and management: The world scenario. *Critical Reviews in Environmental Science and Technology*, *42*(15), 1509–1630. doi:10.1080/10643389.2011.569871

Knox & Andrew. (2005). *An Overview of Incineration and EFW Technology as Applied to the Management of Municipal Solid Waste (MSW)*. University of Western Ontario.

Krishnamoorthy, R. V., & Vajrabhiah, S. N. (1986). Biological activity of earthworm casts: An assessment of plant growth promoter levels in casts. *Proceedings of the National Academy of Sciences of the United States of America*, *95*, 341–351.

Laner, D., Crest, M., Scharff, H., Morris, J. W. F., & Barlaz, M. A. (2012). A review of approaches for the long-term management of municipal solid waste landfills. *Waste Management (New York, N.Y.)*, *32*(3), 498–512. doi:10.1016/j.wasman.2011.11.010 PMID:22188873

Lim, P. N., Wu, T. Y., Sim, E. Y. S., & Lim, S. L. (2011). The potential reuse of soybean husk as feedstock of Eudrilus eugeniae in vermicomposting. *Journal of the Science of Food and Agriculture*, *91*(14), 2637–2642. doi:10.1002/jsfa.4504 PMID:21725978

Lim, S. L., Wu, T. Y., & Clarke, C. (2014). Treatment and biotransformation of highly polluted agro-industrial wastewater from palm oil mill into vermicompost using earthworms. *Journal of Agricultural and Food Chemistry*, *62*(3), 691–698. doi:10.1021/jf404265f PMID:24372356

Lim, S. L., Wu, T. Y., Lim, P. N., & Shak, K. P. Y. (2015). The use of vermicompost in organic farming: Overview, effects on soil and economics. *Journal of the Science of Food and Agriculture*, *95*(6), 1143–1156. doi:10.1002/jsfa.6849 PMID:25130895

Lim, S. L., Wu, T. Y., Lim, P. N., & Shak, K. P. Y. (2015b). The use of vermicompost in organic farming: Overview, effects on soil and economics. *Journal of the Science of Food and Agriculture*, *9*(6), 1143–1156. doi:10.1002/jsfa.6849 PMID:25130895

Lim, S. L., Wu, T. Y., Sim, E. Y. S., Lim, P. N., & Clarke, C. (2012). Biotransformation of rice husk into organic fertilizer through vermicomposting. *Ecological Engineering*, *41*, 60–64. doi:10.1016/j.ecoleng.2012.01.011

López-González, J. A., Suárez-Estrella, F., Vargas-García, M. C., López, M. J., Jurado, M. M., & Moreno, J. (2015). Dynamics of bacterial microbiota during lignocellulosic waste composting: Studies upon its structure, functionality and biodiversity. *Bioresource Technology*, *175*, 405–416. doi:10.1016/j.biortech.2014.10.123 PMID:25459849

Lopez-Mosquera, M. E., Moiron, C., & Carral, E. (2000). Use of dairy industry sludge as fertilizer for grassland in northwest Spain: Heavy metal levels in the soil and plants. *Resources, Conservation and Recycling*, *30*(2), 95–109. doi:10.1016/S0921-3449(00)00058-6

Makan, A., Assobhei, O., & Mountadar, M. (2014). Initial air pressure influence on in-vessel composting for the biodegradable fraction of municipal solid waste in Morocco. *International Journal of Environmental Science and Technology*, *11*(1), 53–58. doi:10.100713762-013-0434-6

Maňáková, B., Kuta, J., Svobodová, M., & Hofman, J. (2014). Effects of combined composting and vermicomposting of waste sludge on arsenic fate and bioavailability. *Journal of Hazardous Materials*, *280*, 544–551. doi:10.1016/j.jhazmat.2014.08.024 PMID:25209831

Marshall, R. E., & Farahbaksh, K. (2013). Systems approaches to integrated solid waste management in developing countries. *Waste Management (New York, N.Y.)*, *33*(4), 988–1003. doi:10.1016/j.wasman.2012.12.023 PMID:23360772

Mc Grath, S., Postma, L., McCormack, R. J., & Dowdall, C. (2000). Analysis of Irish sewage sludges: Suitability of sludge for use in agriculture. *Irish Journal of Agricultural and Food Research*, *39*, 73–78.

Minale, M., & Worku, T. (2014). Anaerobic co-digestion of sanitary wastewater and kitchen solid waste for biogas and fertilizer production under ambient temperature: Waste generated from condominium house. *International Journal of Environmental Science and Technology*, *11*(2), 509–516. doi:10.100713762-013-0255-7

Morgan, J. E., & Morgan, A. J. (1999). The accumulation of metals (Cd, Cu, Pb, Zn and Ca) by two ecologically contrasting earthworm species (*Lumbricus rubellus Aporrectodea caliginosa*): Implications for ecotoxicological testing. *Applied Soil Ecology*, *13*(1), 9–20. doi:10.1016/S0929-1393(99)00012-8

Ndegwa, P. M., & Thompson, S. A. (2001). Integrating composting and vermicomposting in the treatment and bioconversion of biosolids. *Bioresource Technology*, *76*(2), 107–112. doi:10.1016/S0960-8524(00)00104-8 PMID:11131792

Neuhauser, E. F., Loehr, R. C., & Malecki, M. R. (1998). The potential of earthworms for managing sewage sludge. In C. A. Edwards & E. F. Nauhauser (Eds.), *Earthworms in Waste and Environmental Management* (pp. 9–20). The Hague: SPB Academic Publishing.

Page, A. L., Logan, T. J., & Ryan, J. A. (1987). *Land Application of Sludge*. Chelsea, MI: Lewis Publishers.

Paradelo, R., Moldes, A. B., & Barral, M. T. (2013). Evolution of organic matter during mesophilic composting of lignocellulosic winery wastes. *Journal of Environmental Management*, *116*, 18–26. doi:10.1016/j.jenvman.2012.12.001 PMID:23274588

Price, J. S. (1988). Development of a vermicomposting system. Agricultural waste management and environmental protection. In *Proceedings of the 4th CIEC (International Scientific Centre of Fertilisers) Symposium held in Braunschweig*. German Federal Republic.

Price, J. S., & Phillips, V. R. (1990). An improved mechanical separator for removing live worms from worked organic wastes. *Biological Wastes*, *33*(1), 25–37. doi:10.1016/0269-7483(90)90119-D

Qian, X., Shen, G., Wang, Z., Guo, C., Liu, Y., Lei, Z., & Zhang, Z. (2014). Co-composting of livestock manure with rice straw: Characterization and establishment of maturity evaluation system. *Waste Management (New York, N.Y.)*, *34*(2), 530–535. doi:10.1016/j.wasman.2013.10.007 PMID:24188923

Rahuman, A. A. (1984). *Functional Morphology of the Megascolecid Earthworm Lampito mauritii kinberg* (M.Phil., Dissertation). University of Madras, Madras.

Ravindran, B., & Sekaran, G. (2010). Bacterial composting of animal fleshing generated from tannery industries. *Waste Management (New York, N.Y.)*, *30*(12), 2622–2630. doi:10.1016/j.wasman.2010.07.013 PMID:20727727

Sánchez-Monedero, M. A., Serramiá, N., Civantos, C. G., Fernández-Hernández, A., & Roig, A. (2010). Greenhouse gas emissions during composting of two-phase olive mill wastes with different agroindustrial by-products. *Chemosphere*, *81*(1), 18–25. doi:10.1016/j.chemosphere.2010.07.022 PMID:20708773

Sarma, B. K., Singh, P., Pandey, S. K., & Singh, H. B. (2010). Vermicompost as modulator of plant growth and disease suppression. In N. Karmegam (Ed.), *Vermitechnology II* (Vol. 4, pp. 58–66). Dyn. Soil Dyn. Plant.

Seo, S. A., Aramaki, T., Hwang, Y., & Hanaki, K. (2004). Environmental impact of solid waste treatment methods in Korea. *Journal of Environmental Engineering*, *130*(1), 81–89. doi:10.1061/(ASCE)0733-9372(2004)130:1(81)

Shafawati, S. N., & Siddiquee, S. (2013). Composting of oil palm fibres and *Trichoderma* spp. as the biological control agent: A review. *Int. Biodeteriorat. Biodegrad.*, *85*, 243–253. doi:10.1016/j.ibiod.2013.08.005

Sierra, J., Desfontaines, L., Faverial, J., Loranger-Merciris, G., & Boval, M. (2013). Composting and vermicomposting of cattle manure and green wastes under tropical conditions: Carbon and nutrient balances and end-product quality. *Soil Research (Collingwood, Vic.)*, *51*(2), 142–151. doi:10.1071/SR13031

Sigua, G. C. (2005). Current and future outlook of dredged and sewage sludge materials in agriculture and environment. *Journal of Soils and Sediments*, *5*(1), 50–52. doi:10.1065/jss2004.10.113

Sim, E. Y. S., & Wu, T. Y. (2010). The potential reuse of biodegradable municipal solid wastes (MSW) as feedstocks in vermicomposting. *Journal of the Science of Food and Agriculture*, *90*(13), 2153–2162. doi:10.1002/jsfa.4127 PMID:20718020

Singh, A., & Sharma, S. (2002). Composting of a crop residue through treatment with micro-organisms and subsequent vermicomposting. *Bioresource Technology*, *85*(2), 107–111. doi:10.1016/S0960-8524(02)00095-0 PMID:12227532

Singh, J., & Kalamdhad, A. S. (2014). Potential for composting of green phumdi biomass of Loktak lake. *Ecological Engineering*, *67*, 119–126. doi:10.1016/j.ecoleng.2014.03.086

Singh, R. P., & Agrawal, M. (2009). Use of sewage sludge as fertilizer supplement for *Abelmoschus esculentus* plants: Physiological, biochemical and growth responses. *International Journal of Environment and Waste Management*, *3*(1/2), 91–106. doi:10.1504/IJEWM.2009.024702

Singh, R. P., & Agrawal, M. (2010a). Effects of different sewage sludge amendment rates on growth, biomass accumulation, nutritional quality, heavy metal accumulation and yield of mung bean Vigna radiata L. plants. *Ecological Engineering*, *36*, 969–972. doi:10.1016/j.ecoleng.2010.03.008

Singh, R. P., & Agrawal, M. (2010b). Variations in heavy metal accumulation, growth and yield of rice plants grown at different sewage sludge amendment rates. *Ecotoxicology and Environmental Safety*, *73*(4), 632–641. doi:10.1016/j.ecoenv.2010.01.020 PMID:20163857

Singh, R. P., Singh, P., Araujo, A. S. F., Ibrahim, M. H., & Sulaiman, O. (2011). Management of urban solid waste: Vermicomposting a sustainable option. *Resources, Conservation and Recycling*, *55*(7), 719–729. doi:10.1016/j.resconrec.2011.02.005

Soobhany, N., Mohee, R., & Garg, V. K. (2015). Comparative assessment of heavy metal content during the composting and vermicomposting of municipal solid waste employing Eudrilus eugeniae. *Waste Management (New York, N.Y.)*, *39*, 130–145. doi:10.1016/j.wasman.2015.02.003 PMID:25708402

Suthar, S. (2009). Vermicomposting of vegetable market solid waste using Eisenia foetida: Impact of bulking material on earthworm growth and decomposition rate. *Ecological Engineering*, *35*(5), 914–920. doi:10.1016/j.ecoleng.2008.12.019

Suthar, S. (2014). Toxicity of methyl parathion on growth and reproduction of three ecologically different tropical earthworms. *International Journal of Environmental Science and Technology*, *11*(1), 191–198. doi:10.100713762-012-0154-3

Tang, Z., Yu, G., Liu, D., Xu, D., & Shen, Q. (2011). Different analysis techniques for fluorescence excitation-emission matrix spectroscopy to assess compost maturity. *Chemosphere*, *82*(8), 1202–1208. doi:10.1016/j.chemosphere.2010.11.032 PMID:21129765

Tian, Y., Chen, L., Gao, L., Michael, F. C. Jr, Keener, H. M., Klingman, M., & Dick, W. A. (2012). Composting of waste paint sludge containing melamine resin and the compost's effect on vegetable growth and soil water quality. *Journal of Hazardous Materials*, *243*, 28–36. doi:10.1016/j.jhazmat.2012.09.013 PMID:23127275

Tomati, U., Galli, E., Pasetti, L., & Volterra, E. (1995). Bioremediation of olive mill waste waters by composting. *Waste Management & Research*, *13*(6), 509–518. doi:10.1177/0734242X9501300602

Wang, J., Hu, Z., Xu, X., Jiang, X., Zheng, B., Liu, X., ... Kardol, P. (2014). Emissions of ammonia and greenhouse gases during combined pre-composting and vermicomposting of duck manure. *Waste Management (New York, N.Y.)*, *34*(8), 1546–1552. doi:10.1016/j.wasman.2014.04.010 PMID:24835490

Wang, Q. R., Cui, Y. S., Liu, X. M., Dong, Y. T., & Christrie, P. (2003). Soil contamination and plant uptake of heavy metals at polluted sites in China. *Journal of Environmental Science and Health. Part A, Environmental Science and Engineering & Toxic and Hazardous Substance Control*, *38*, 823–838. PMID:12744435

Watteau, F., & Villemin, G. (2011). Characterization of organic matter microstructure dynamics during co-composting of sewage sludge, barks and green waste. *Bioresource Technology*, *102*(19), 9313–9317. doi:10.1016/j.biortech.2011.07.022 PMID:21807502

Wu, T. Y., Lim, S. L., Lim, P. N., & Shak, K. P. Y. (2014). Biotransformation of biodegradable solid wastes into organic fertilizers using composting or/and vermicomposting. *Chemical Engineering Transactions*, *39*, 1579–1584.

Yadav, A., & Garg, V. K. (2011). Industrial wastes and sludges management by vermicomposting. *Reviews in Environmental Science and Biotechnology*, *10*(3), 243–276. doi:10.100711157-011-9242-y

Yadav, K. D., Tare, V., & Ahammed, M. M. (2010). Vermicomposting of source-separated human faeces for nutrient recycling. *Waste Management (New York, N.Y.)*, *30*(1), 50–56. doi:10.1016/j.wasman.2009.09.034 PMID:19850460

Yadav, K. D., Tare, V., & Ahammed, M. M. (2012). Integrated composting-vermicomposting process for stabilization of human faecal slurry. *Ecological Engineering*, *47*, 24–29. doi:10.1016/j.ecoleng.2012.06.039

Chapter 4
Modeling and Prediction of Zone of Critical Concern:
A Solid Waste Management Case of a Higher Education Institute in India Using Probit Regression

Jishnu Bhattacharyya
https://orcid.org/0000-0002-2691-6629
Indian Institute of Information Technology and Management, Gwalior, India

Ben Krishna
https://orcid.org/0000-0003-1591-5492
Indian Institute of Management, Kozhikode, India

Mukesh Narmetta
Indian Institute of Management, Kozhikode, India

Olea Roy
Jadavpur University, Kolkata, India

Soumyadeep Kundu
KIIT University, Bhubaneswar, India

ABSTRACT

The on-campus residence and activities of students, faculty members, visitors, and university staff result in the generation of university campus solid waste (UCSW). An accurate forecast of UCSW generation is crucial for proper on-campus waste management. This chapter, therefore, aims at identifying significant variables that affect the excess production of waste at a higher education institute in Kozhikode in the Indian state of Kerala. The study is expected to improve the waste management

DOI: 10.4018/978-1-7998-1966-0.ch004

program therein. A literature review methodology, along with a series of interviews, has been used to frame and execute a questionnaire-based study where waste management staff provides the required responses. The modeling technique used is a Probit regression. The influential variables that affect the excess generation of waste are average education level, the status of employment, the urban background of most residents, consumption of water and electricity, space understudy, and management control. All the results in this research can be utilized as a part of waste management programs and in improving capacities to manage waste. This will also help in understanding the type of areas that are zones of critical concern in terms of waste management. The observations incorporated in the study is in the Indian context and are unique among the existing literature.

INTRODUCTION

Since the dawn of human settlements and residential living areas, solid waste has been a major concern. The practice of municipal waste management can be traced back to the People of Crete, Greece's largest island, in about 1500 BC. The citizens used to practice dumping the rubbish in large pits and covered it with dirt. One of the earliest laws concerning an organized form of municipal waste management may be seen in the history of Athens, Greece in about 500 BC. The law requires the dumping of trash at least a mile out of residential living area/town, and they were not allowed to dump on the streets. During the middle ages, the problem of solid waste was common across European cities, and the severity touched its maxima in 1349 with the Black Death/Bubonic plagues. The plague reduced half of Europe's total population. Beyond health, in the 1400s, the mountain of garbage just outside the city of Paris in France interfered with the city's defense arrangements. Then in 1893, German used technology to manage the solid waste with its first solid waste incinerator. With time new technologies, laws, and regulatory policies were introduced to manage solid waste in cities and other parts of the country (Uriarte, 2008). The practice and concerns related to solid waste can even be seen in ancient Asian civilization like the current day Republic of India or the former Hindustan and various manuscripts like the ancient literature of Kautilya"s Arthshastra (The Maurya's period). In Arthshastra, it was stated that that maintaining sanitation of habitat was essential and inevitable for a good quality of life. The Hindu society was aware of the code of conduct or Dharma relevant to waste management in the residential areas/Kingdom, and cremation of dead human bodies was one of such practice (Sharma and Upadhyay, 2009; Gottlieb, 2003). Even an elaborate sanitary and drainage system was a characteristic of ancient Indus Valley Civilizations (Drains, 2019).

The wastes are generated from residential, industrial, commercial, institutional, and agricultural areas as also from construction sites, hospitals, and treatment plants. They comprise primarily of solid materials that have served their purpose and are therefore fit for disposal. In the present times, urban centers have become hubs of solid waste generation, inflating the probability of creation of polluted and unhygienic environments that facilitate the spread of vector-borne and other diseases. Dumping of wastes in the open will, for example, allow the biodegradable waste materials to decompose naturally, producing foul smells and multiplying disease-causing germs. Toxic wastes from industries can change the biological and chemical properties of soil while seeping into the ground to poison groundwater reserves. Burning of hazardous wastes along with scrap articles like paper can produce gases that are capable of causing severe diseases in humans and animals. These are all, therefore, improper methods of solid waste management. Solid waste management is a major challenge in urban areas throughout the world, and an effective and efficient waste management program is necessary to avoid health hazards and negative impact on the environment (Krishna, Manickam, Shah, and Davergave, 2017). A number of technologies and methods have been developed and applied for waste management in the past few years and there are volume of research addressing the cause (Hazardous Waste Management - an overview, 2019; Karmperis, Aravossis, Tatsiopoulos, and Sotirchos, 2013; Allesch and Brunner, 2014; Cobo, Dominguez-Ramos, and Irabien, 2018; Coban, Ertis, and Cavdaroglu, 2018).

The practice of efficient solid waste management deals with the collection, treatment, and disposal of solid wastes while also paying heed to the recycling of suitable materials. Some of the popular methods include the operation of sanitary landfills, incineration, composting, pyrolysis, and recycling. However, owing to the rapid growth of population and consumerism in the developing world, proper management of solid wastes therein has become a recurring issue. In India itself, as of 2009, 90 percent of solid wastes were disposed of in open, unsanitary landfills (Narayanan, 2009). In one of the major Indian city Bengaluru, an estimated 20 lakh households generate around 4,200 tons of solid waste each day, and most of it is being transported to unscientifically planned landfills (The Hindu, 2018). Further, less than 60% of waste is collected from Indian households, and only 15% of the total urban waste is processed. Hyper consumption is a curse of our modern times, and we are continuing to generate monumental amounts of waste (Sankar, 2019).

For addressing this issue, characterization of the wastes being generated, and the available methods of their collection, segregation, and disposal are required to be studied. With the practice of sustainability, gaining popularity and encouragement worldwide, it is simultaneously logical to move beyond studying only the waste management practices of municipalities, to exploring the waste management systems at educational institutions. This study aims to do likewise, focusing on a premier business school located in Kozhikode, Kerala.

Unanticipated levels of human activity, among other factors, lead to the generation of excessive solid wastes. The objective of this paper is to model and predict the production of excess wastes on the campus of the Kozhikode Business School[1], in the process detecting the variables that significantly affect the possibility of excessive waste generation. Fifteen variables are hypothesized to be affecting the probability of excess waste generation, which is the dependent variable. A Probit model is used to capture the relationship between the fifteen independent variables and excess waste generation. Some of these independent variables are the number of people living on campus, the number of visitors, the average age group of residents, their average educational qualifications, the temperature of the environment, consumption levels of water and electricity, and others. However, finally, only six of the fifteen factors that were initially identified are found to significantly affect the generation of excess solid waste on campus. Convenient sampling was used with eighty sample points. Interviews were conducted with the waste management staff at the institute, and vital spaces on the campus were continuously observed. The methodology chosen finds support from a round table discussion among researchers (18 researchers who are well-versed with multiple research methods) from across management and engineering disciplines.

The research, henceforth, is organized as follows. After a brief review of literature that addresses the waste management practices in India and around the world, the observations for the same in the case of IIMK are recorded. Subsequently, the methodology for modeling is explicitly discussed. A summary of the results obtained follows, along with a discussion of the findings and a conclusion. The appendices bear certain observations made and the questionnaire used for conducting the interviews.

LITERATURE REVIEW

Waste Management

Extant studies discussing waste management practices focus on four processes: waste generation, waste collection, waste composition, and waste disposal. In most countries, Government bodies like Municipalities are responsible for managing these processes. Waste managed by municipal corporations are commonly referred to as Municipal Solid Waste (MSW), and it is crucial to study these processes for gaining a clear understanding of the waste management practices of an area or a municipality.

Definition of Municipal Solid Waste

Municipal waste is collected and treated by municipalities. It covers waste from households, including bulky waste, waste from commerce and trade, office buildings, institutions, small businesses, yard and garden, street sweepings, contents of litter containers, and market/public place cleansing. Waste from municipal sewage networks and treatment, as well as municipal construction and demolition, are excluded.

Municipal solid waste (MSW) (also called trash) consists of everyday items such as product packaging, yard trimmings, furniture, clothing, bottles and cans, food, newspapers, appliances, electronics, and batteries. Sources of MSW include a residential waste (including waste from multi-family housing) and waste from commercial and institutional locations, such as businesses, schools, and hospitals. The Environmental Protection Agency's (EPA) definition of MSW does not include industrial, hazardous, or construction and demolition (C&D) waste. Once generated, MSW must be collected and managed. Common management methods include recycling, composting, combustion with energy recovery, and landfilling. Many wastes that are landfilled represent a loss of materials that could be reused, recycled, or converted to energy to displace the use of virgin materials. Source: United States Environmental Protection Agency

Sources and Types of MSW

MSW contains waste from various sources such as residential waste (food waste, paper, cardboard, plastic, textile, consumer electronics), industrial waste such as (packaging, housekeeping waste, construction, demolition material, hazardous wastes) commercial waste (paper, cardboard, plastic, wood, glass, metal, e-waste), institutional waste (same as commercial waste), construction and demolition (wood, steel, concrete, dirt, bricks, tiles), and municipal services (street sweeping, landscape, tree trimming, waste from parks and other areas (Hoornweg & Bhada, 2012).

Waste Management Practices Around the World

Worldwide, 1.3 billion tons of solid waste is generated annually, which is expected to go up to 2.2 billion by 2025. The OECD[2] countries produce 44% of the total waste (Esposito, Tse, and Soufani, 2016).

A study by World Bank (2012) states "As the world hurtles toward its urban future, the amount of MSW, one of the most important by-products of an urban lifestyle, is growing even faster than the rate of urbanization. Ten years ago, there were 2.9 billion urban residents who generated about 0.64 kg of MSW per person per day (0.68 billion tonnes per year). This report estimates that today these amounts have increased to about 3 billion residents generating 1.2 kg per person per day (1.3 billion tonnes per year). By 2025 this will likely increase to 4.3 billion urban residents generating about 1.42 kg/capita/day of municipal solid waste (2.2 billion tonnes per year)|| (Hoornweg & Bhada, 2012).

Main methods to dispose of waste adopted worldwide are landfills, recycling, dumps, composts, incineration, and others. The most commonly used method of disposing of waste across the world is the landfill. In the OECD, more than 60 percent of waste is diverted from landfills while in Africa, most of the collected waste is dumped (World Bank Report, 2012). In developing countries, informal recycling by waste pickers at the collection points recover a significant portion of discards (Hoornweg et al., 2005). In the US, the government has adopted a pricing mechanism: a way to reduce waste generation and increase recycling. Under the pricing mechanism, product charges can stimulate consumer behavior for waste reduction as they need to pay for waste management services while purchasing the products. One more pricing policy is that a consumer who separates recyclable waste would pay less than those who don‘t.

Waste Management Practices in India

Due to rapid industrial growth, population growth and urbanization the problem of solid waste management has become severe in developing countries like India, China, etc., efficient waste management requires good knowledge of the waste that is being generated and also requires expertise on methods of collection and disposal of the waste (Narayana, 2009). Often it is observed that the municipal corporations of India do not have the expertise and resources necessary to handle the current situation. ——Waste management and disposal are a pressing issue facing India today since about 90% of waste is currently disposed of by open dumping (Narayana, 2009).

There are many scientific methods available for efficient handling of solid wastes across the globe of which Indian city municipalities can adopt few that are appropriate for their needs. ——The two leading innovative mechanisms of waste disposal being adopted in India include composting (aerobic composting and vermicomposting) and waste-to-energy (WTE) (incineration, palletization, bio-methanation). WTE projects for disposal of MSW are a relatively new concept in India (Sharholy, Ahmad, Mahmood, and Trivedi, 2008).

The various methods available for handling solid waste are as follows:

1. Landfilling (Sharholy et al., 2008)

Though there are many solutions available for proper solid waste management, many cities adopt uncontrolled, poorly managed, and open dumping processes, resulting in degradation of the environment. Dumping activities in the coastal cities have increased the heavy metals being injected into the waters (Sharholy et al., 2008). In most of the cities across India, solid waste is being dumped in low-lying areas outside city premises without following any sanitary landfilling principles or techniques although the city municipality is aware of the fact that Sanitary dumping is the only acceptable and recommended method of disposal (Sharholy et al., 2008). ——Uncontrolled dumping of wastes on outskirts of towns and cities has created overflowing landfills, which are not only impossible to reclaim because of the haphazard manner of dumping, but also have serious environmental implications in terms of groundwater pollution and contribution to global warming (Gupta, Mohan, Prasad, Gupta, and Kansal, 1998). Leaving the dumped waste unattended will decompose naturally that give rise to bad odors, hosts, and feeds insects and pests, which leads to severe health problems (Sharholy et al., 2008).

2. Recycling of organic waste (Sharholy et al., 2008)
    a. Aerobic composting: Aerobic Composting is the method of bacterial conversion of organics present in the solid waste in the presence of air under hot and moist conditions. The final product of this process is of high agricultural value and is called humus or compost. This can be used as a natural fertilizer (Sharholy et al., 2008). This method was encouraged by the Govt. of India through its fourth Five Year Plan (1969-1974) by providing grants and soft loans to the state governments to set up composting plants across various urban cities of India.
    b. Vermicomposting: In this method, organic waste is stabilized by the joint action of earthworms and aerobic microorganisms. The earthworms feed on this solid waste.

The end product can be used as a biofertilizer in agriculture. ——If the organic waste is buried in pits under partially anaerobic conditions, it will be acted upon by anaerobic microorganisms with the release of methane and carbon dioxide; the organic residue left is good manure (Sharholy et al., 2008).

3. Thermal treatment techniques (Sharholy et al., 2008)
    a. Incineration (Narayana, 2009): In this process, solid waste is burnt under a controlled environment. It helps recover energy and destruct solid waste such as waste from hospitals. Incineration reduces the waste by 80 to 90 percent of its original volume. This makes incineration the most

attractive, but unfortunately, it is not practiced widely in India (Sharholy et al., 2008)

b. Gasification techniques: Gasification is the process of incineration in which solid waste is subjected to oxygen-deficient conditions. This process can be attractive as it can help reduce pollution while managing solid waste (Sharholy et al., 2008).

c. RDF plants: ——The main purpose of the refuse-derived fuel (RDF) method is to produce an improved solid fuel or pellets from MSW. The RDF production is about 210 t/day as fluff and pellets, and it is going to be used for producing power (about 6.6 MW)(Sharholy et al., 2008). Subsequently, RDF plants can reduce the pressure on landfills.

4. Recovery of recyclable materials (Sharholy et al., 2008)

Apart from the elements that can be composted, a large number of materials such as paper, plastic, ferrous and non-ferrous material, rubber, etc., present in the solid waste can be recovered and recycled. An estimated 13-20% of the solid waste contains recyclable material across various cities in India (Sharholy et al., 2008). Compared to the government, the private sector is more interested in collecting this recyclable material, and it is minimal compared to the informal sectors such as rag pickers (Sharholy et al., 2008).

Waste Management Practices at Kozhikode Business School: The Case

There are a total of 250 collection points on the campus of which some need daily attention (such as classrooms and library points). All the hostels have collection points at various levels such as individual rooms to floors to one in each hostel. The housekeeping staff has a pre-defined schedule for collection of the waste depending upon the priority of the place (based on their perception). There are 40 to 50 employees to take care of sanitation needs on the campus. The maximum working hours for each of these employees is 6 hours in practice. The limited employee capacity is managed by serving different points on different days, and high attention areas like classrooms are served every day. There is weekly job rotation practice (within sanitation service). The whole campus produces around 100 – 200 kg of solid waste per day (the amount of waste generated doubles during festivals and events that are occasionally arranged on campus) including food waste in the canteen (excluding three major private restaurants that are completely managed by the leased partners). There are three major canteens run by the institution through private catering agency.

The campus is home to more than 1000 students, faculties, and staff members. The whole campus is divided into the academic block and student residential block in one hill and staff residential block in a separate hill. This case deals with the academic block, student residential block, and all common areas on the campus excluding staff residential area, which is managed separately. The campus is in the coordinates: 11°18′23″N 75°52′44″E. And is characterized by hilly terrain. As such it is difficult for the sanitation staff to move from one point to the other point as they either need to climb up/down hundreds of stairs or need to take long ways in the stiff terrain (worse during long monsoon season). This result is inefficiency and underutilization of capacity.

As a result of these unscientific and under planned arrangements the major problems is that there is not enough staff for daily collection of waste generated which leaves few collection points vulnerable to untidiness and the management often fail to anticipate the amount of waste getting generated at various point and faces capacity problem. Further, the management fails to allocate limited human resource with proper capacity across various collection points on the campus. It was often noted by the observer that the sanitation staff is not aware of the amount of waste that may be generated across various points and often need to serve the same point twice. Again, it is observed that they are not aware of the points that can have more generated waste and need priority service. It was further noted that the installed waste collection capacity at various pints are often inadequate for the serviced area.

The campus estate manager and infrastructure committee sought a long term solution to this problem. They call for a proposal that may help them understand the campus waste generation pattern and help them to predict excess waste generation points on campus. The proposal in the form of the project was given to a senior professor in the institute who specializes in operations and analytics. The project was finally given to a group of scholars specializing in various domain of management. The solution was a Probit regression-based model, as discussed further.

There is no doubt that centralized waste management models are suitable for this case, but there is a lacuna in managing it completely. This is primarily because of the lack of information about waste generation across various service points on the campus. As in the case mentioned in this study do not have any process in place for their sanitary waste generation capacity prediction. Thus, end up going for an arbitrary assignment of the workforce. This brings attention to the need for developing a predictive model.

Waste Management and Modeling

For a predictive model of waste generation, many methods were used in literature. Few of them are multivariate analysis, AI systems (Artificial Neural Network), and multiple regression methods. Noori (2012) used the artificial neural network, and multiple linear regression analysis (Wei, Xue, Yin and Ni, 2013), Mwenda (2014) used time series analysis. In 2011, Li et al. (Li et al., 2011) used the analytical hierarchy process. Choice of modeling method depended mainly on the type of data. For cross-sectional data, the multiple linear regression approach was used more frequently using socio-economic variables (Kannangara, Dua, Ahmadi, and Bensebaa, 2018).

Study Area (Observation and Data Collection Units)

In literature, areas considered for the study of prediction modeling of Municipal Solid Waste (MSW) were mainly household units, residential units (urban and rural), municipal regions, districts and in some studies country-wise data was also collected and used for modeling.

Independent Variables

Many independent variables were used in past studies for predictive modeling of MSW generation. Most common independent variables used were population, GDP or income, age, education, and the consumption level of residents (Wei, Xue, Yin and Ni, 2013). Apart from these, the temperature was also considered as an independent variable in past studies (Keser et al., 2012), as were urbanization, seasonal variables, and consumption expenditure (Kolekar, Hazra, and Chakrabarty, 2016).

RESEARCH METHODOLOGY

In this paper, we explore the case study of a premier business school in the state of Kerala in the Republic of India and the adjoining area of the campus under the institute's control. There have been numerous attempts made by the students, staffs, and faculties to make an effort to manage their waste and particularly solid waste in a centralized manner. The sampling of the research has a case study that accounts for one academic block, one student residential block, and common campus area with various collection points that houses near 1000 people; the student residential block is a collection of multiple hostels, and the academic block is a single integrated unit (that have classrooms, staffrooms, faculty cabins, security rooms, seminar halls, small food outlets, meeting rooms, library, Guest house for executives, and exhibition rooms). Each of them is lessons on waste management endeavors and proved to be the basis of this research.

The study is based on a literature review, and a questionnaire was administered based on the literature review and interview with waste management staffs. In the study, excess waste generation is evaluated by checking the unit of study's allocated disposal capacity. If the evaluators find that the capacity provided is not sufficient and is overflowing, then the evaluator assigns ——Excess Waste Generation. There were 4 evaluators who have assigned scores independent of each other at three different time of the day and then the final score came after detailed discussion until consensus is reached. All forms of disagreements were resolved through discussion mode. A 5th person was responsible to manage the four evaluators and monitor the research process. He was the lead in the discussion and responsible for bringing consensus. The 5th person was aware of activates undertaken by the four evaluators and was continuously observing them. He was responsible for keeping up with integrity and lay standard of the research. The disposal capacity is provided in every public and private space in the Kozhikode Business School campus of similar capacity. Individual spaces on the Kozhikode Business School campus are observed like the FPM Office, Library, Student Store, Café Seating Area, etc. and these individual spaces came up with individual observations. All the observations were collected offline and were then converted into electronic format. The data were analyzed using the open-source tool, Gretl. Further, the Probit model is used to predict excess waste generation.

In the standard settings like the multivariate measurement on some random cross-sectional data or repeated measurements on a sample of subjects across time, correlated binary data arises. The model formulation is the central issue because of such data (Chib, 1998). ——One strategy, outlined by Carey, Zeger, and Diggle (1993) and Glonek and McCullagh (1995), relies on the generalization of the binary logistic model to multivariate outcomes in conjunction with a particular parameterized representation for the correlations (Chib, 1998). Another strategy is the use of a binary Probit model. ——The resulting multivariate Probit model is described in terms of a correlated Gaussian distribution for underlying latent variables that are manifested as discrete variables through a threshold specification (Chib, 1998).

Sample Size and Hypothesis

The calculation of the sample size is based on the power and the effective size. In practice, many prediction studies, considers 5% as significance level, 80% statistical power, and R2 value as 0.25(Wong, 2013). Also, we consider odd ratio as 3.5, considering the outcome concerning the exposure of variables to be high (Odd ratio >1) increased the occurrence of an event. Thus we estimated the sample size and found to be 62 as compared to our sample size 80. So, the sample size is adequate for the desired research outcome.

The test of correlation of the variables found out those variables are weakly correlated, which implies that it will not decrease the statistical power as it will not increase standard error of estimation. The ten-per-variable rule is considered

Table 1. Design of the survey

Research type	Exploratory
Sampling technique	Convenient sampling
Sampling units	Individual spaces of Kozhikode Business School
	The campus is observed like the FPM
	Office, Library, Student Store, Café
	Seating Area, etc.
Sample Size	80
Tools for data collection	Questionnaire
The medium of data collection	Offline
Tools for data analysis	Gretl

Figure 1. Correlation matrix

	Noofpe~y	Noofvi~s	Aveage~p	Aveedu~l	status~t	backgr~t	consum~c	consup~s	ocassi~e
Noofpeople~y	1.0000								
Noofvisitors	0.0868	1.0000							
Aveagegrp	-0.0261	-0.0502	1.0000						
Aveeduclevel	0.1146	0.0678	-0.3735	1.0000					
statusofempt	0.3308	0.1209	-0.0725	0.5386	1.0000				
background~t	-0.2690	-0.2023	-0.0575	0.1390	-0.1081	1.0000			
consumptio~c	0.1629	0.1119	-0.1753	0.5664	0.6374	-0.0710	1.0000		
consuphabits	0.1816	-0.1291	0.0766	0.2267	0.5200	0.1826	0.1841	1.0000	
ocassional~e	0.0063	-0.0310	0.4975	-0.6402	-0.4245	-0.0492	-0.5185	-0.0450	1.0000
spaceunder~y	-0.1126	-0.0888	-0.1236	0.1559	0.0429	0.1689	-0.0532	0.0608	-0.2162
directlyma~K	0.0186	0.0422	0.2239	-0.1216	-0.0553	0.2490	-0.0055	0.1616	0.2548
Toomanythi~e	-0.0570	0.0564	-0.0291	0.1034	-0.1070	-0.0780	0.0931	-0.2669	-0.0526
wellmainta~d	-0.0676	0.1108	-0.0180	0.1089	0.0172	-0.2547	0.0345	-0.3553	0.0625

	spaceu~y	direct~K	Tooman~e	wellma~d
spaceunder~y	1.0000			
directlyma~K	0.0723	1.0000		
Toomanythi~e	-0.0700	-0.2650	1.0000	
wellmainta~d	-0.1367	-0.4449	0.2204	1.0000

a rule of thumb, and in our case, certain relaxation is applicable as the number of the observations are restricted to the number of room availability in the campus. There are various studies which do not follow the rule of thumb, and one of the most cited works is Vittinghoff and McCulloch (2007), which explains 5-9 events per predictor variables, is sufficient to a decent prediction. In our case, we have 15 variables and approximately 5-6 events per predictor variable and is adequate for the selected sample size.

MULTIVARIATE MODELING HYPOTHESES

Predictive Model for Excess Waste Generation

This model explains the probability of Excess Waste Generation based on the observed variables. The hypothesis related to each factor is explained below. These hypotheses are based on the literature review and a result of discussion with waste management officials at Kozhikode Business School.

H1: Expense Level increases *Excess Waste Generation* probability.
H2: A higher number of people staying will increase *Excess Waste Generation* probability
(Ojeda-Benitez et al. 2008)
H3: Increase in the number of visitors will increase *Excess Waste Generation* probability
H4: Increase in the average age group will increase *Excess Waste Generation* probability
(Ojeda-Benitez et al. 2008)
H5: Increase in average education level will increase *Excess Waste Generation* probability
(Ojeda-Benitez et al. 2008)
H6: Increase in the status of employment will increase *Excess Waste Generation* probability (Lebersorger and Beigl, 2011
H7: Urban background of more residents will increase *Excess Waste Generation* probability (Lebersorger and Beigl, 2011)
H8: Increase in consumption of water and/or electricity will increase *Excess Waste Generation* probability
(Liu and Yu, 2007)
H9: Non-green consumption habits will increase *Excess Waste Generation* probability (Dai et al. 2011).
H10: 'Occasionally, the number of visitors varies **will** increase *Excess Waste Generation* probability
(Abbasi et al. 2012)
H11: Space understudy will increase *Excess Waste Generation* probability
H12: Directly managed my IIMK will increase *Excess Waste Generation* probability
H13: Increase in Temperature will increase *Excess Waste Generation* probability (Keser et al., 2012).
H14: Too many things in the space will increase *Excess Waste Generation* probability
H15: Well-maintained and decorated ambiance will increase *Excess Waste Generation* probability

The above hypothesis is emanated from the extensive literature review (available in the reference section) conducted and is also based on the series of interviews conducted with the staffs involved in waste disposal and management in the higher education institute under study. The hypothesis 1, 4, 5, 6, 7 relate to the socio-economic strata the individuals belong to since it is evident from the literature review and also through our interviews that as a person has a higher income and moves to a higher socio-economic strata, their expenses on consumable items will increase, resulting in an increase of waste generation (Ojeda-Benitez et al. 2008). If we consider the case of the United States and India, USA has comparatively less population than India, but they belong to a comparatively higher socio-economic stratum than regular Indians and generate more waste than that of a regular Indian. The hypotheses 8, 9 correspond to the nature of the individual and his/her attitude towards the environment. The hypotheses correspond to the sustainable behavior exhibited by the individual in the form of the nature and frequency of consumption and their disposable practices (Liu and Yu, 2007 & Dai et al. 2011). The hypotheses 2, 3, 10, 11, 14 correspond to the more quantitative aspects of waste generation. As may be expected, an area having a higher population density compared to an area having a lower population density for the same socio-economic factors will have a higher waste generation due to the higher population. The same can be applied for the area under consideration, which corresponds both the acreage under consideration as well as the kind of area being studied. This was corroborated by the interviews of the housekeeping staff who stated that the waste generated from student residences having a higher number of students generated more waste than that of the ones having lower student residents. They also mentioned how waste was more in case of student residences than that of the academic block where regular teaching takes place. Since the population has an effect on the amount of waste generated, with the influx of more people, there would be a higher waste generation (Abbasi et al. 2012). In the case of the studied institution, it was corroborated by the cleaning staff who mentioned the increase in waste generated during conferences and other events on campus which see the influx of people. The hypotheses 12, 13, and 15 correspond to the ecological and institutional aspects. According to Keser et al. (2012), they mention that an increase in temperature will lead to increase in waste generation. The study found this to be applicable in the case of the institute under study as well and hence included it considering that it is located in a place which experiences high swings in temperature. Hypothesis 12 and 15 was generated following interviews of the housekeeping staff who mentioned the high amount of waste generated in the upkeep of the campus as mandated by the institute's administration. Since the administration will have a hand in the regular upkeep of the institute, and also the various events organized, it will have an impact on the waste generated in the process, hence warranting the introduction of hypothesis 12. The justifications for the hypotheses are drawn from the logical implications of the investigation, which was a test to ascertain their significance.

The probability that the unit of study will Generate Excess Waste is represented by the following Probit model.

$P(Excess = 1/X) = F(Z_i)$

where $F(Z_i)$ is the cumulative normal distribution given by

$$F(Z_i) = \frac{1}{\sigma\sqrt{2\pi}} e^{-\frac{1}{2}\left(\frac{Z_i^2}{\sigma}\right)}$$

where,

$Z_i = \beta_0 + \beta_1$* (Expense Level) + β_2* (Number of people staying) + β_3* (Number of Visitors) + β_4*(Average Age Group) + β_5*(Average Education Level) + β_6*(Status of employment) + β_7*(Urban background of most resident) + β_8*(Consumption of water and/or electricity) + β_9*(Consumption habits) + β_{10}*(Occasionally Number of Visitor Varies) + β_{11}*(Space under study) + β_{12}*(Directly managed my IIMK) + β_{13}*(Temperature) + β_{14}*(Too many things in the space) + β_{15}*(Well maintained and decorated) + e

RESULTS AND DISCUSSION

Predictive Model for Excess Waste Generation

*Only the significant values are expressed in the table. Insignificant values are either discarded or not considered in the actual model to define the causal relationship.

Thus, the model is explained by

$P(Excess=1/X) = F(Z_i)$,

where $F(Z_i)$ is the cumulative normal distribution given by

$$F(Z_i) = \frac{1}{\sigma\sqrt{2\pi}} e^{-\frac{1}{2}} \left(\frac{Zi^2}{\sigma}\right),$$

Table 2. Results Obtained

Explanatory Variables	coefficient	Std. Error	Z-value	p-value	Slope
Average Education Level***	2.34845	0.848153	2.769	0.0056	0.759624
Status of employment***	−4.04289	1.23085	−3.285	0.0010	−0.954395
background of most resident***	−2.03310	0.683413	−2.975	0.0029	−0.507219
Consumption of water and/or electricity**	−0.925099	0.428308	−2.160	0.0308	−0.349552
Space under study*	0.663594	0.391156	1.696	0.0898	0.250742
Directly managed my IIMK**	−2.06648	1.01277	−2.040	0.0413	−0.562786

***significant at 1%,**significant at 5%,*significant at 10%

$Z_i = \beta_5$*(Average Education Level) + β_6*(Status of employment) + β_7*(Urban background of most resident) + β_8*(Consumption of water and/or electricity) + β_{11}*(Space under study) + β_{12}*(Directly managed my IIMK) + e

The estimation results indicate that most of the variables that have been identified from literature tend to have an association with the dependent variable, Generate Excess Waste'. When the education level of the resident moves from 0 to 1 (low to high), holding other variables at their mean, the change in probability from no excess to excess waste is -0.759. Similarly, an increase in the status of employment decreases the probability of excess waste accumulating by 95.4 percent. It is interesting to observe that urban background, and more consumption of water/electricity decreases the excess waste generation probability. When the space under study, switches from open to private, the probability of excess waste increases by 25.04 percent. If IIMK directly manages the space under study, then the probability of excess waste generation decreases by 56.27 percent. Thus, it is recommended that space is directly managed by IIMK. The status of the hypotheses can be summarized as follows:

Table 3. Summary Status of the Hypotheses

Hypothesis	Significance	Comments
H1	Not Significant	
H2	Not Significant	
H3	Not Significant	
H4	Not Significant	
H5	Significant	Hypothesis validated as in literature
H6	Significant	Hypothesis from literature not valid in this case. And produce opposite results
H7	Significant	Hypothesis from literature not valid in this case. And produce opposite results
H9	Significant	Hypothesis from literature not valid in this case. And produce opposite results
H10	Not Significant	
H11	Significant	Hypothesis from literature not valid in this case. And produce opposite results
H12	Significant	Hypothesis from literature not valid in this case. And produce opposite results.
H13	Not Significant	
H14	Not Significant	
H15	Not Significant	

CONCLUSION

With matters of solid waste management gaining importance in an increasingly globalized and industrialized world, the paper had set out to explore solid waste management practices at an educational institution of a developing country, India. The objective was to identify the variables that significantly affect the probability of generation of excess waste at a premier business school in Kozhikode, Kerala. The study adopted an exploratory nature. A review of literature proved crucial in selecting the research methodology, framing the questionnaire, and outlining the variables under study. Accordingly, a Probit model was used to identify the relationships between Excess Waste Generation' a dependent variable, and fifteen independent variables. The findings rejected eight of the independent variables based on insignificance (p-value greater than 0.05).

In the final model, the average education level of occupants is found to be negatively related to excess waste generation, which is expected. As the person becomes educated, he/she becomes more conscious of the environment, leading to a lower generation of wastes. Similarly, movement from a lower to a higher status of employment reduces the probability of generation of excess waste. If the residents/visitors are from an urban background or consume more water and electricity, the probability of generation of excess waste once again reduces. On the contrary, if the space under study changes from open or public to private, the probability of excess waste generation rises. This hints at the presence of an interesting phenomenon that makes individuals more comfortable with generating more waste at private places. However, if the space under study is directly managed by the host institute, the probability of excess waste generation immediately drops by 56.27 percent. Thus, a recommendation of the study would be for the institute to personally supervise or control the waste management practices in the spaces on the campus where waste is likely to be generated. The model generated in this paper can be useful for predicting and preparing for efficient solid waste management at the institute under study. The methodology adopted lends itself to replicability in other studies that may, in the future, choose to research excess waste generation at different educational institutes.

Limitations

This piece of research case was limited by available resources, and with this brings with it a few limitations that may be improved upon in practice and future research. All universities generate solid waste, and India has many hundreds of universities, deemed universities, private engineering colleges, and various autonomous institutes, etc. The idea that the results from this research can produce a model for many or all institutions is implausible in its current form and therefore the study must be replicated in various higher education institutes all across the country and if possible globally or at least among SAARC nations. This will give us a larger sample size and will better represent the target population. Then we may get a model that is more generalizable across higher education institutions. Solid waste is almost universal as a problem in India and generally is regarded as unacceptable by Westerners (In personal communication by Dr. M. R. Brett-Crowther). Therefore, getting samples beyond the geographic limitations of the Republic of India is highly desired. The observations need to be compared and contrasted from the study to the case of other institutes/universities in India and globally. In future studies, one may also consider the failure of the institute/university or the State government to apply the country's existing anti-pollution laws.

Further, adding analysis that local corruption or negligence or incompetence was a causal factor, could add to the existing body of knowledge significantly. In future research, the cultural aspects of the geography under study and their influence on the higher education institutes may bring some interesting insights. Extending further, if future research can incorporate insights related to treatment that the sanitation workers get from the residents, and the consequence may bring new understanding in the domain.

REFERENCES

Allesch, A., & Brunner, P. H. (2014). Assessment methods for solid waste management: A literature review. *Waste Management & Research*, *32*(6), 461–473. doi:10.1177/0734242X14535653 PMID:24895080

Coban, A., Ertis, I. F., & Cavdaroglu, N. A. (2018). Municipal solid waste management via multi-criteria decision-making methods: A case study in Istanbul, Turkey. *Journal of Cleaner Production*, *180*, 159–167. doi:10.1016/j.jclepro.2018.01.130

Cobo, S., Dominguez-Ramos, A., & Irabien, A. (2018). From linear to circular integrated waste management systems: A review of methodological approaches. *Resources, Conservation and Recycling*, *135*, 279–295. doi:10.1016/j.resconrec.2017.08.003

Dai, C., Li, Y. P., & Huang, G. H. (2011). A two-stage support- vector-regression optimization model for municipal solid waste management- A case study of Beijing, China. *Journal of Environmental Management*, *92*(12), 3023–3037. doi:10.1016/j.jenvman.2011.06.038 PMID:21872384

Esposito, M., Tse, T., & Soufani, K. (2016). Companies are working with consumers to reduce waste. *Harvard Business Review*.

Gottlieb, R. S. (2003). *This sacred earth: Religion, nature, environment* (2nd ed.). New York: Routledge. doi:10.4324/9780203426982

Gupta, S., Mohan, K., Prasad, R., Gupta, S., & Kansal, A. (1998). Solid waste management in India: Options and opportunities. *Resources, Conservation and Recycling*, *24*(2), 137–154. doi:10.1016/S0921-3449(98)00033-0

Harappa. (2019). *Drains*. Available at: https://www.harappa.com/category/slide-subject/drains

Hoornweg, D., & Bhada-Tata, P. (2012). *What a waste: a global review of solid waste management 15*. Washington, DC: World Bank.

Kannangara, M., Dua, R., Ahmadi, L., & Bensebaa, F. (2018). Modeling and prediction of regional municipal solid waste generation and diversion in Canada using machine learning approaches. *Waste Management, 74*.

Karmperis, A. C., Aravossis, K., Tatsiopoulos, I. P., & Sotirchos, A. (2013). Decision support models for solid waste management: Review and game-theoretic approaches. *Waste Management, 33*(5), 1290-1301.

Keser, S., Duzgun, S., & Aksoy, A. (2012). Application of spatial and non-spatial data analysis in determination of the factors that impact municipal solid waste generation rates in Turkey. *Waste Management (New York, N.Y.)*, *32*(3), 359–371. doi:10.1016/j.wasman.2011.10.017 PMID:22104614

Kolekar, K., Hazra, T., & Chakrabarty, S. (2016). A Review on Prediction of Municipal Solid Waste Generation Models. *Procedia Environmental Sciences*, *35*, 238–244. doi:10.1016/j.proenv.2016.07.087

Kollikkathara, N., Feng, H., & Yu, D. (2010). A system dynamic modeling approach for evaluating municipal solid waste generation, landfill capacity and related cost management issues. *Waste Management (New York, N.Y.)*, *30*(11), 2194–2203. doi:10.1016/j.wasman.2010.05.012 PMID:20547450

Krishna, I. M., Manickam, V., Shah, A., & Davergave, N. (2017). *Environmental management: science and engineering for industry* (1st ed.). Oxford, UK: Butterworth-Heinemann.

Lebersorger, S., & Beigl, P. (2011). Municipal solid waste generation on municipalities: Quantifying impacts of household structure, commercial waste and domestic fuel. *Waste Management (New York, N.Y.)*, *31*(9-10), 1907–1915. doi:10.1016/j.wasman.2011.05.016 PMID:21689921

Lebersorger, S., Schneider, F., & Hauer, W. (2003). Waste generation in households — models in theory and practical experience from a case study of multifamily dwellings in Vienna. *Proceedings of the Ninth International Waste Management and Landfill Symposium.*

Martin, E. J., Chawla, R. C., & Swartzbaugh, J. T. (2004) Hazardous waste site remediation technology selection. Waste Management Series, 4, 1019-1066. doi:10.1016/S0713-2743(04)80040-0

Narayana, T. (2009). Municipal solid waste management in India: From waste disposal to recovery of resources? *Waste Management, 29*(3), 1163-1166.

Pires, A., Martinho, G., & Chang, N. B. (2011). Solid waste management in European countries: A review of systems analysis techniques. *Journal of Environmental Management*, *92*(4), 1033–1050. doi:10.1016/j.jenvman.2010.11.024 PMID:21194829

Sankar, M. (2019). *Cleaning up the mess: the need for a waste management policy.* Available at: https://www.thehindu.com/opinion/op-ed/cleaning-up-the-mess/article26763338.ece

ScienceDirect. (2019). *Hazardous Waste Management-an overview, ScienceDirect Topics*. Available at: https://www.sciencedirect.com/topics/earth-and-planetary-sciences/hazardous-waste-management

Sharholy, M., Ahmad, K., Mahmood, G., & Trivedi, R. C. (2008). Municipal solid waste management in Indian cities – A review. *Waste Management (New York, N.Y.)*, *28*(2), 459–467. doi:10.1016/j.wasman.2007.02.008 PMID:17433664

Sharma, A., & Upadhyay, V. (2009). *Environment and Ecology (Web): Module 1*. Available at: https://nptel.ac.in/courses/122102006/mod1/1.htm

The Hindu. (2018). *The challenge of solid waste management*. Available at: https://www.thehindu.com/life-and-style/homes-and-gardens/the-challenge-of-solid-waste-management/article25205083.ece

United States Environmental Protection Agency. (n.d.). *Municipal Solid Waste*. Available at: https://cfpub.epa.gov/roe/indicator_pdf.cfm?i=53(accessed 12 August 2019)

Uriarte, F. A. (2008). Solid Waste Management: Principles and Practices: an Introduction to the Basic Functional Elements of Solid Waste Management, with Special Emphasis on the Needs of Developing Countries. Quezon City: UP Press.

Vittinghoff, E., & McCulloch, C. E. (2007). Relaxing the rule of ten events per variable in logistic and Cox regression. *American Journal of Epidemiology*, *165*(6), 710–718. doi:10.1093/aje/kwk052 PMID:17182981

Wei, Y., Xue, Y., Yin, J., & Ni, W. (2013). Prediction Of Municipal Solid Waste Generation In China By Multiple Linear Regression Method1. *International Journal of Computers and Applications*, *35*(3). doi:10.2316/Journal.202.2013.3.202-3898

Wong, K. K. K. (2013). Partial least squares structural equation modeling (PLS-SEM) techniques using SmartPLS. *Marketing Bulletin*, *24*(1), 1–32.

ENDNOTES

1 The name of the institute is hidden for privacy purpose.

2 The Organisation for Economic Co-operation and Development is an intergovernmental economic organisation with 36 member countries, founded in 1961 to stimulate economic progress and world trade.

APPENDIX 1

Few observations used in the study (A Demonstration)

Waste Management Facilities at Kozhikode Business School
Waste Bins
Waste Bins

APPENDIX 2

Few observations used in the study (Demonstration)

Dumping Yard at Kozhikode Business School

APPENDIX 3

Few observations used in the study (Demonstration)

Satellite View of Kozhikode Business School
Dumping Yard at Kozhikode Business School (Satellite View)
Dumping Yard was once green at Kozhikode Business School (Satellite View-2014)
Dumping Yard near residential area (Satellite View)

APPENDIX 4

Design of questionnaires

DEPENDENT VARIABLES

Questionnaire	0	1
	Limited Waste Generated	Excess Waste Generation*
Outcome as justified by evaluators		
*Excess waste generation is evaluated by checking the unit's alocated disposal capacity. If the evaluators find that the capacity provided is not sufficient and is overflowing that the evaluator assign "Ecess Waste Generation". The disposal capacity is provided in every public and private space in the Kozhikode Business School campus based on possible waste generation.		

INDEPENDENT VARIABLES: Part 1

Questionnaire	0	1	2	
	Low (<8K)	Medium(8-15K)	High (15K +)	
Expense Level?				Economic Indicator
	1-10	**10-20**	**20+**	
Number of people staying? (Ojeda-Benitez et al. 2008)				
	1-10	**10-20**	**20+**	
Number of Visitors?				
	20-30	**30-40**	**40+**	
Average Age Group? (Ojeda-Benitez et al. 2008)				Population Indicators
	<UG	**UG**	**PG**	
Average Education Level? (Ojeda-Benitez et al. 2008)				
	Contractual		**Full Time**	
Status of employment (Lebersorger and Beigl, 2011;)				

INDEPENDENT VARIABLES: Part 2

Questionnaire	0	1	2	
	Rural		Urban	
Urban background of most resident? (Lebersorger and Beigl, 2011)				Population Indicators
	Less	**Average**	**Excess**	
Consumption of water and/or electricity? (Liu and Yu, 2007)				
	Green Practices		**Non Green practices**	Consumption Indicators
Consumption habits(Dai et al. 2011).				
	No		**Yes**	
Occasionally Number of Visitor Varies (Abbasi et al. 2012),				**Seasonal Indicators**
	Open Space	**Private**	**Office**	
Space under study				
	No		**Yes**	Legal Indicator
Directly managed my IIMK				

INDEPENDENT VARIABLES: Part 3

Questionnaire	0	1	2	
	Low	Medium	High	
Temperature (Keser et al., 2012).				Ambience Indicator
	No		**Yes**	
Too many things in the space				
	No		**Yes**	
Well maintained and decorated				

APPENDIX 5

Variables description

Exp_Lev= Expense level of the individual
No_people= Number of people staying
No_visitors= Number of visitors
Avg_age= Average age group
Avg_edu= Average Education Level (Matriculation, Higher Secondary, Graduate, Post-Graduate)
Emp_stat= Employment Status (High income employment, Middle Income employment, Low Income, employment Unemployed)
Urban= Urban background of residents (City, Town, Semi-Urban, Rural)
Util_con= Consumption of electricity and water
Con_habit= Consumption Habit (High, Medium, Low)
Var_visitors= Occasionally number of visitor varying
Space= Space under study i.e. student residence, academic block, administrative block
Dir_man= Directly managed by IIM Kozhikode i.e. whether there is direct involvement of administration in the process
Temp= Increase in temperature
Th_space= Too many things in space
Main= Status of maintenance and upkeep

Chapter 5
Sustainable Development Model and Innovation of Mineral Resources

Song Jiang
Xi'an University of Architecture and Technology, China

ABSTRACT

Mineral resources are important material resources for the development of national economy, but the long-term linear and extensive mining mode leads to low exploitation and utilization of resources and serious waste of resources on the one hand; on the other hand, the environment of mining areas is deteriorating, and safety problems are prominent. Correctly handling the relationship between mineral resources, environment, and economic development has become the main problem of sustainable development of mineral enterprises. There are still some problems in the mine. Therefore, it is particularly necessary to explore the construction path model of green mine under the background of the new era. Based on the coordinated development of resources, environment, and regional economy, this chapter studies the green mine construction mode by using the knowledge of resource economics, mining science, and sustainable development theory, combining theory with empirical analysis and qualitative and quantitative analysis.

DOI: 10.4018/978-1-7998-1966-0.ch005

INTRODUCTION

Mineral resources are the material basis for economic development and play a very important role in the development of China's economy. However, according to the current situation in China, there are many problems in the process of resource development and utilization in China. Nowadays, China is in the stage of rapid development of science and technology. In order to implement the concept of sustainable development, China is adopting high-tech technology to continuously transform resources to reduce unnecessary losses(Drielsma,2016).Previously, China's mining resources mining technology is still at a low level, and the level of productivity is not high, which has caused the efficiency of mining mineral resources in China to be particularly low, making China a serious waste of mineral resources for a long time(Zhang,2017). To this end, China is striving to find efficient mining models and environmentally friendly mine management utilization models in response to the concept of sustainable development(Lei,2016). The work of Niekerk (2015) has been used by many professors to focus on how to inject the concept of sustainable development into the innovation of the mine management model. For the time being, China is still in a period of shortage of resources and energy, and the problem of endless exploitation and waste of mineral resources is also beginning to emerge. In order to meet the growing demand for minerals with the development of society, miners can only dig a large amount of mineral resources continuously, but because the mining technology adopted in China is still in a relatively backward stage, resulting in a large amount of waste, so that China can not continue A large number of mining can not meet the dilemma of development needs. At this time, The authors of Rossi (2012), Moosavi (2015) and Wang(2016) discuss in their research the world is holding high the banner of sustainable development of resources, and mining companies are thus aware that the old mine management model and backward mining technology can no longer meet the current development needs, and innovation management is imminent. Therefore, enterprises should implement the concept of sustainable development in the improvement of mining technology and the innovation of mine management mode, and take an environmentally friendly resource utilization road.

At present, China has discovered more than 160 kinds of minerals, and it has almost all the minerals in the world. Among them, there are more than 140 kinds of proven reserves, which are far ahead in the world's mineral resources. However, China has a large population base and a small per capita possession, and there are many problems in the development and utilization of mineral resources, resulting in a shortage of resources due to the consumption of a large amount of mineral resources. The shortage of resources and environmental pollution have seriously affected the sustainable development of human society, and the contradiction between the exploitation and utilization of mineral resources and sustainable economic development has become increasingly prominent(Liu, 2016). Moreover, China has the goal of "building a well-off society in an all-round way". In order to achieve this goal, China cannot rely too much on the inherent way of developing economic resources by consuming mineral resources. We should implement the scientific development concept, change the economic development mode according to China's national conditions, and follow the new industrialization road of China to meet the major challenges of sustainable development of natural resources and economy.

STATUS OF DEVELOPMENT AND UTILIZATION OF MINERAL RESOURCES

In recent years, due to the weak economic recovery and the sluggish demand, the global mining situation continues to be sluggish. Domestic mining enterprises are carrying forward the burden under the dual test of China's new economic normality and the new mining situation(Vollenbroek,2002). Most mining enterprises have poor natural endowment conditions, inflexible institutional mechanisms, heavy debt burden, high cost, and the traditional thinking of mine production and management. Under the influence of factors, profitability has fallen sharply, the loss is getting bigger and bigger, and even the capital chain is broken. It can be said to be on the verge of survival. With the introduction of the new economic normal, China's economic growth has slowed down, and the consumption demand of the tertiary industry has gradually become the main body, gradually shifting from investment-driven to innovation-driven. Especially in 2017, China has entered the year of major reform and major adjustment. With the "supply-side reform" boosting the development of the economy, it has pointed out the direction and focus for future macroeconomic reforms.

In the economic restructuring and transformation and upgrading, the focus is on compressing excess capacity and accelerating the elimination of backward production capacity in the steel and coal industries. The world's three major iron mineral resources mining companies are making every effort to expand production, accelerate technological transformation and new technology applications, production efficiency continues to increase, production and operating costs are gradually reduced, forcing high-cost iron ore suppliers to gradually bankrupt and close the market, the future The three major mining enterprises will continue their dominant position in the international iron ore shipping market, the market share will continue to increase, and the market monopoly will continue to increase(Carayannis,2017). As far as domestic iron ore mining enterprises are concerned, most of them are characterized by small scale, low grade, high cost and high debt. The market competitiveness and survivability are far lower than those of major international mining companies. China's metallurgical mining enterprises currently have the following crises:

1. The disorderly expansion of steel production capacity has caused blind construction of mines. China's steel production capacity accounts for 50% of the world's total production capacity. Over the past 10 years, due to the sharp increase in the price of mineral products, China's exploration rights and mining rights have been expanding, and some mining enterprises have been blindly put into operation, and some low grades and small reserves have been built. Inferior mines with high costs. In 2016, there were 28,425 registered exploration rights in the country, with a registered area of 549,745 km2; 74,023 registered mining rights, a registered area of 104,083 km2, and a production scale of more than 15 billion yuan.
2. The operating performance has been seriously declining, and the survival and development are difficult. Since 2013, mining companies have entered the cold winter period, and prices have fallen in a cliff-like manner. Starting in February 2016, iron ore prices rebounded as the steel market turned warmer. At present, the common problems in mining enterprises are high cost, limited space for cost reduction, low return on assets, income of main business of 45.6 yuan per 100 yuan of assets, 43% of the average level of industrial enterprises, and low per capita income. .
3. High-cost, low-capacity enterprises have withdrawn from the market and become zombie enterprises. A group of serious enterprises with losses suffered once stopped production, but the withdrawal mechanism of enterprises has not been established, and the exit policy has not been established. Often, after the capital chain breaks, enterprises still cannot retreat from the market and wait for the market to recover.
4. Enterprise funds are difficult, personnel are aging, and transformation is difficult. When the industrial economy is in recession and enterprises are facing difficulties, banks will often reduce or reduce loans in order to reduce bad debts. Under such circumstances, enterprises will face the risk of capital chain breakage.
5. The burden of taxation, safety and environmental protection of enterprises is heavy. In July 2016, the State Administration of Taxation stipulated that the resource tax was changed from the original quantity to the ad valorem. After the reform, some enterprise resource taxes were reduced, but some mining enterprises' resource tax increased by one time; Expenditure on expenses is relatively large. In order to reduce the burden on enterprises that use the mining method, the state stipulates that the resource tax shall be reduced by half; the mining enterprises in the closed production period shall be reduced by 30%.

However, in the specific implementation, some companies did not enjoy this offer because of the cumbersome work. In recent years, with the country's control of safety and environmental protection, enterprises' investment in safety and environmental protection equipment and losses caused by limited production will bring a heavy burden to enterprises.

6. The impact of imported iron ore from abroad. China's self-produced iron ore accounts for only about 15% of domestic consumption, and most of the rest comes from imports. Domestic iron ore enterprises are difficult to compete with imported iron ore. International mining giants manipulate the market and have a serious impact on China's iron ore industry and even the steel industry. They use the low-cost dumping strategy to fight the domestic iron ore industry.

China's metallurgical mining industry is struggling under various crises. However, the big waves and sands, the mines that can survive this global mining winter, will be broken into butterflies and Nirvana. Therefore, the key now is to find a way for the metallurgical mining enterprises to maintain a stable and stable development to avoid the risks caused by the international and domestic environment. The key to the survival of enterprises is economic benefits. Therefore, it is necessary to study the price mechanism and cost of China's mineral products, especially the factors affecting the price of mineral products and the external costs that are easily ignored by enterprises, and predict the price trend in advance. Recognize the cost component of mineral products and reasonably avoid risks. In the face of irreversible international environment, only by changing itself can we avoid risks and maintain the healthy survival and development of the company. As far as the production management level of mining enterprises is concerned, China's mining enterprises, especially metallurgical mining enterprises, lag far behind other industries, facing the needs of the international environment and the new era, using management model innovation to improve the production management level of mining enterprises and promote It is the survival of metallurgical mining enterprises that mine enterprises reduce costs and increase efficiency and realize the transformation and upgrading of mining enterprises. In the process of resource-based enterprise interest evaluation mechanism and behavior change, the development model of resource-based industry also changes, as shown in Figure 1. Therefore, the sustainable development model is an inevitable choice driven by value-based industries.

As the environment changes, the company's interest evaluation mechanism is also changing, and constantly adjust its behavior. Due to the increasingly urgent resource and environmental issues, the concept of sustainable development has been generally accepted by the state and society. The evaluation of the interests of resource-based enterprises has shown dynamic changes, and its behavior has also shown obvious stages(Babie,2013). Stage 1: Maximize production. Under

Figure 1. Evolution of the development model of resource industry under the change of corporate interest evaluation mechanism

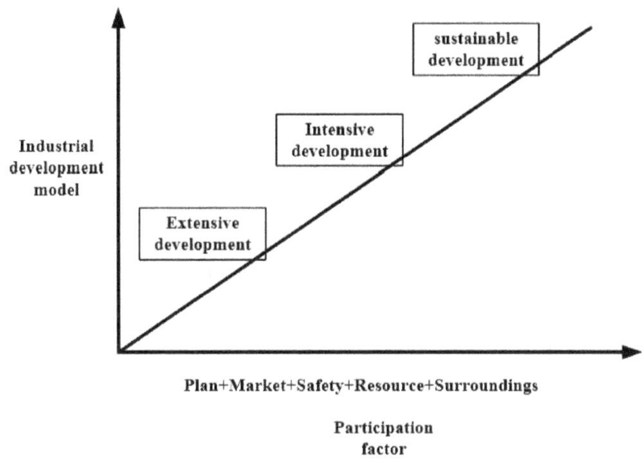

the planned economic conditions, the resource-based enterprises are owned by the government, the products are purchased and sold, and the enterprises are not facing the market. The enterprise's goal is to complete the production tasks assigned by the state. In this context, resource-based enterprises operate extensively, waste resources, and cause serious environmental damage. Phase 2: Maximizing economic benefits. After the reform and opening up, enterprises must face the market, enterprises become independent economic entities, and profits become the basic goal pursued by enterprises. Enterprises should not only consider how to expand production, but also consider how to reduce costs. Similarly, in this context, resource and environmental issues and security issues remain prominent. Phase 3: Maximizing overall benefits. At the end of the 1990s, due to the seriousness of China's resource and environmental problems and the implementation of national sustainable strategies, resource-based enterprises are facing more and more resource and environmental constraints, resource-based enterprises' resource conservation, environmental protection, and safe production. The impact of corporate benefits is growing. Under such a background, enterprises began to pursue the maximization of comprehensive benefits, pay attention to the comprehensive utilization of resources and environmental protection during the exploitation and utilization of resources, so that China's resources and environment problems have improved.

PROBLEMS FACING THE SUSTAINABLE DEVELOPMENT OF MINERAL RESOURCES

In the context of sustainable development, the development and management of domestic mineral resources is severely restricted. Halme(2014), Liu(2008) and Sharfman (2000) showed in their previous work that it is necessary to ensure the maximum benefit of economic resources and meet the existing funding policy requirements. Only through the revision of mineral resources development and management can we Ensure the sustainable and sustainable development of mineral resources. However, with the implementation of the policy, many problems and obstacles are encountered in the development and management of mineral resources. For this reason, this paper analyzes the problems in the development and management of mineral resources, and provides feasible solutions for the development and management of mineral resources through analysis of problems.

Mineral Resources Development Management System is not Perfect

China's mining administrative management system is imperfect, forming a management system of "segmentation" at the central, provincial, municipal, and county levels. The problem of mine ownership in some areas has not been resolved, and the property rights of the mines have not been clarified. Many government departments and enterprises have been exploiting. They sell minerals for economic benefits, completely ignoring the impact on the environment, earning black money, drilling legal loopholes, breaking mining order, regardless of whether mineral resources are suitable for mining. Nor did it go through surveys and use scientific means to conduct measurement and demonstration, but blindly open mining and build factories in the mining area. Some areas are still unrestricted, collectively mined, and even some are privately mined under the banner of the state. Without strict control of the system, the behavior of some areas is not standardized, which ultimately affects the sustainable development of the economy.

In the context of sustainable development, China's mineral resources development management has not yet formed a complete system. Many mineral resources development and management departments only carry out quantitative management according to existing relevant laws and regulations, and the management of mineral resources development is relatively general. The lack of effectiveness and lack of a special management system, the existing mineral resources development management system is not only seriously lagging behind and is contrary to the sustainable development strategy. In view of the current state of mineral resources development in China, combined with the development needs of the market economy and the policy of sustainable development, the mineral resources development management system is optimized, and the original mineral resources development management audit mode is adjusted to realize the development of mineral resources. Dynamic supervision, using modern information means, timely release of mineral resources development and management, refine the legal system, and implement sustainable development policies. Establish a responsibility system in the mineral resources development management system, strictly control the operation mechanism, reference content, implementation methods and efficiency in the management system, and promote the implementation of sustainable development policies.

The Management Method of Mineral Resources Development is Unreasonable

China is one of the major resource countries with rich mineral resources and a wide variety of mineral resources in the world. Among them, the per capita possession of rare earth ore, tungsten ore and tin ore is high. According to relevant information, China has found more than 160 kinds of minerals, including energy minerals, metal minerals, non-metallic minerals and other major categories. There are more than 140 proven reserves. The total amount is large, and it has been proven to account for 13% of the world, ranking third in the world, with the United States and Russia. However, due to the large population of China, the per capita possession is small(Gassman,2005). It is only 59% of the world's per capita possession and ranks 54th in the world, far behind other countries. The existence of this problem seriously restricts the sustainable development of China's economy. Mineral resources are used in industrial production to produce industrial waste. According to relevant data, the comprehensive utilization rate of industrial waste in China is only 28%, and the utilization efficiency is low, causing serious industrial pollution. Moreover, it is also a waste of resources. Mineral waste leads to the dependence of China's resources on a large number of imports. Increasing environmental problems not only threaten people's lives, but also hinder the sustainable development of the economy. The non-implementation of national environmental protection work has led to a negative impact on the environment during the development and utilization of mineral resources. The irrational development of mineral resources has led to a sharp decline in forests and more and more serious desertification. The irrational use of mineral resources has led to an increasingly obvious greenhouse effect. It will not only pollute the local environment, but also endanger the sustainable development of society.

The central government proposes a strategy of sustainable development that is inconsistent with the economic development goals of local governments. At the level of mineral resources development and management strategy, the central government pays attention to the sustainable development of mineral resources and promotes the comprehensive development and sustainable use of mineral resources. Some minerals are reasonably developed and managed. However, local governments shoulder the goal of economic development, especially in areas where mineral resources are the economic support. Through the full development of mineral resources to achieve economic growth, although the economic growth rate is relatively significant, but the environment and mineral resources are over exploited, It has already caused the emergence of some resources to dry up. It shows that the development and management of mineral resources is inconsistent with the existing economic policies and contradicts the strategy of the central government for sustainable development. Due to the different resource enrichment in each region, the overall development planning management is also different, but the development planning content of mineral resources is almost the same, which leads to the actual content of guiding actual development and management, due to the mineral resources. Different storage volumes but the same management methods will inevitably bring resistance to the development and management of mineral resources, resulting in many local areas not within the scope of planning and coordination, and privately developing mineral resources. The original mineral resources development and management method has continued the original management mode. The quality of planning management is not high, and it does not have information technology. Eventually, management and execution are separated in the link. There is no timely information feedback as the basis for management. Mineral resources Development management is not efficient. The function of mineral resources development and management is relatively vague, and there is no clear functional synergy, which makes the management work very complicated. From the central government to the local government, the speed and implementation speed of the policy is seriously lagging behind, in the actual development and management of mineral resources. In the middle, the management institutions established are not strict with the management of actual minerals, and the responsibility of other regulatory authorities is insufficient, that is, the consumption of financial expenses does not reflect the functions of the departments, and the budgetary cost of management is increased.

Lack of Professional Talents and the Supervision Mechanism of the Masses

The management of mineral resources development is usually managed by specific departments and lacks public monitoring mechanisms. With incomplete statistics, the average number of professional employees in the development and management of mineral resources is about 400 in each province, but the area and affairs to be managed exceeds the number of people who are responsible for affairs five times, and the level of management personnel is uneven. %, undergraduate or above accounted for 17%, indicating that some employees have not received good management training. The supervision mechanism for public participation is seriously backward. There is no special petition mechanism. The masses lack the supervision of mineral resources development and management. It is difficult for practitioners to absorb the suggestions provided by the masses. The establishment of the supervision mechanism can improve the efficiency of mineral resources development and management, balance the resource development management process through the supervision of departments, appropriately introduce modern information technology, improve the rigor of the supervision mechanism through technical means, and improve the feedback awareness of the management process.

Core Objectives are not Fully Realized

Sustainable development is an important decision for the benefit of future generations. The departments of mineral resources development and management should take policy as the core and actively implement the purpose and function of the policy. Most local government management departments compare the implementation of policies. Delays can not play the meaning of the policy itself, and managers do not fully understand the sustainable development policy. Subjective consciousness still stays in system construction and income generation. The focus of the local government is relatively clear. The original development focus is changed through the policy release. Due to the excessive number of personnel involved, it is difficult to make timely adjustments. The synergy between departments and departments is not high, it is difficult to take policy implementation as the main body, and the departments excuse each other's responsibilities, resulting in a difficult situation after the implementation of the policy. It is necessary for the government to coordinate and integrate relevant management departments to ensure that the sustainable development policies can be implemented smoothly. Obstacles that hinder the implementation of sustainable development policies are not only in the functional departments, but also in local private mining. The policy of sustainable development severely restricts the way private minerals are mined and damages the economic interests of some people. Revenue will deliberately undermine the implementation of sustainable development policies. In response to such problems, relevant departments should actively respond to national policies, and legal protection policies can be implemented.

MANAGEMENT MODEL OF SUSTAINABLE DEVELOPMENT OF MINERAL RESOURCES

The sustainable development of mineral resource-intensive areas is based on the regional mineral resource endowment. While achieving regional economic growth, it maintains and improves environmental quality, creates employment and wealth, enhances regional competitive advantages, and obtains economic, social and environmental aspects. Overall performance improvement. Drawing on the theory of business networks and the theory of national innovation ecosystem, we can construct an ecological innovation system model and hope to achieve the goal of sustainable development of mineral resources. The framework of the pattern is shown in Figure 2.

In the ecological innovation system model framework map of sustainable development in mineral resource-intensive areas, the regional mineral resource endowment is only an important condition for the sustainable development of mineral resources industry function. A systematic change in the culture, values, and habits of a country can promote the formation and evolution of the national innovation system, thereby directly adjusting the positioning of mineral resources in the country's social and economic life, and providing support or constraints for

Figure 2. Framework of sustainable development management model for mineral resources

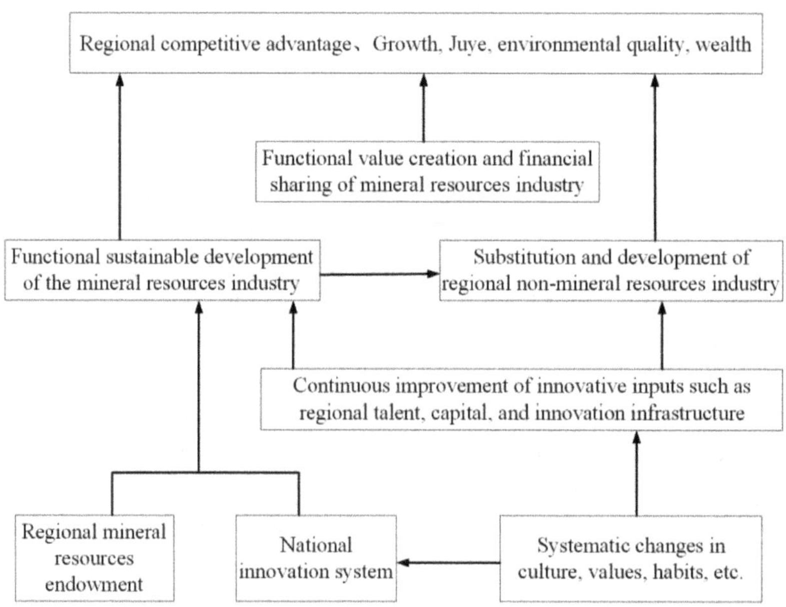

the functional development of mineral resources. It can also promote continuous improvement of innovative inputs such as regional talent, capital, and innovation infrastructure. The continuous improvement of regional innovation investment can lead to changes in the positioning of the mineral resources industry in the regional function. For example, in extreme cases, the mining and mining of the regional mineral resources industry may disappear due to the depletion of mineral resources, and the dissipated may be the function of the region as a resource supply, but the region may be due to mineral processing, mineral resource by-product utilization, The development of mineral equipment, mineral futures trading, mineral resources industry personnel training, consulting, or mining tourism has been successful. The new positioning may not be related to the mineral resources endowment of the region, but only some mineral resources industry derivatives in the region. Alternative functional activities. The improvement of the quality and quantity of regional innovation investment will directly increase the innovation and value of different products and services in the mineral resources industry value chain, and make the development of regional mineral resources industry out of the dilemma of relying solely on resource endowment development, which is conducive to the realization of regional economy. The sustainability of growth and the improvement of environmental quality. On the other hand, the continuous development of the mineral resources industry function can promote the development of related industries through the extension and development of the industrial chain, increase the accumulation of talents, capital and innovation infrastructure and the improvement of the quality of innovative elements, and indirectly support regional non-mineral resources. The development of industry and its replacement of mineral resources industry in regional social and economic activities. In this way, after a period of time, it is possible to achieve a weakening of mineral-intensive areas or even completely rid of the dependence on the mineral resources industry and achieve regional sustainable development. Obviously, it is only expected that through the continuous development of the mineral resources industry function and the replacement and development of the mineral resources industry by the regional non-mineral resources industry, regional sustainable development cannot be achieved because of the continuous development of regional mineral resources industry functions and non-mineral resources. The development of industry itself requires the guarantee of sustainable development conditions. This guarantee is based on the formation of innovative systems and wealth-driven development.

China's original mining enterprise management model usually adopts extensive management mode and method, which is not conducive to the long-term development of China's mining enterprises. Under the new era, with the continuous advancement of China's economic and industrial transformation and upgrading, the management model of mining enterprises also needs to be reformed and innovated. It is necessary to introduce modern management theories and means, and use information platform to upgrade and optimize mining enterprise management. The benign development of mining enterprises.

Streamlined and Efficient Corporate Culture

On the basis of organizational structure, human resource management, organizational culture and symbolic architecture theory, combined with the mine reality, following the principles of streamlining, unification and efficiency, a systematic organizational structure system was constructed. The main features and innovations are reflected in:

The concept of the spirit is clarified. Fuquan Mining expresses the vague corporate spirit and culture concept in a short and clear language, so that every employee can keep in mind and always use it to motivate himself. Such as corporate vision: to become a domestic first-class, internationally renowned metallurgical underground mining enterprises. Enterprise mission: sustainable development and utilization of mineral resources. Values: Performance first, harmony and efficiency; action-oriented, pragmatic and innovative; people-oriented, stressing solidarity and mutual assistance. Entrepreneurial spirit: hard work, environmental protection, excellence, innovation. Service concept: Honesty is attitude, trustworthiness is commitment. Business philosophy: safety, efficiency, harmony and innovation. R&D philosophy: taking pragmatism as the foundation and innovation as the driving force. Employing the concept: caring for people, respecting people, cultivating people, and achieving people. Cost concept: Intensive investment and fine control. Safety concept: Life is wealth, and safety is the family. The overall goal: to maintain sustainable development, to do industry benchmarks.

The reduction of staff and efficiency is normalized in the management of personnel, the implementation of "simplified and efficient, multiple posts, multiple jobs, one multi-functional", strict control of job quotas and personnel recruitment. Each post set in each department has a clear job description, which stipulates the basic conditions, basic skills, business connections and job responsibilities within and outside the company, and has clear responsibilities and clear division of labor.

The flattening of the organizational structure is based on the principle of streamlining, unification and efficiency. It closely focuses on rationalizing the relationship of duties and effectively implementing the flat management mode of the two-level structure--the large-scale system to achieve the organic integration of functions and the original organization from the traditional three. Level management system, namely: company → department → workshop, by changing the workshop setting, the management level becomes the second level, namely: company →

department. The company's new organization is set up as a six-part one, namely: the General Department, the Finance Department, the Safety and Environmental Protection Department (Safety Department, the Ministry of Environmental Protection), the Supply and Marketing Department, the Selection Department, the Transportation Department, and the Production Technology Center.

The performance cost linkage will link the performance appraisal with the production technology and economic indicators such as output, quality and equipment, and at the same time, the office and low-value consumables will be included in the assessment. Performance appraisal management is divided into two levels: organization to organization, organization to individual. That is, the company is on the workshop (department), workshop (department), team or individual. The formulation and implementation of the former assessment plan is specifically undertaken by the company's human resources department; the latter is determined by the workshop (department) according to economic responsibility, and its implementation and implementation are supervised and inspected by the human resources department.

Personnel configuration optimization In the recruitment of personnel, a variety of recruitment channels are adopted, and the entire recruitment and staffing process forms a cycle. Firstly, according to the demand, the application for supplement is proposed, and then the recruitment is carried out. After the recruitment, the existing personnel are optimized again. For the surplus operation and maintenance employees, the "dismissal and voluntary withdrawal mechanism" is adopted to implement the reduction and selection of personnel, so as to continuously optimize the composition of the members. Let employees take the initiative to find their own way. When the company has good efficiency and lack of manpower, it can be re-employed first, and finally achieve a win-win effect.

Party and government work is detailed in the construction of political culture, rationally set up party work institutions, and formulate annual work plans for party building, trade unions, and league committees. Taking the integrity of the industry as the focus, taking the "party mass line education and practice activities" as an opportunity, actively promote the transformation of leading cadres' work style, do practical things for employees, and do good things. In order to ensure the high-quality and efficient development of party building work, design regular meeting system, regular reporting system, party building research system, party building contact point system and establishment of supervision and inspection system. In terms of organizational leadership, the first is to clearly manage leadership. The deputy general manager of the company is in charge of the work of the trade union and directly gives guidance to the work of building a family. Second, the staff and cadres of the trade unions were carefully equipped, and the women workers' committee and the fund review committee were established, which achieved a pattern of rational allocation of resources and joint efforts of the party and government organizations.

The symbol concept deepens the products produced by the enterprise and the services provided is the result of the production and operation of the enterprise. It is the primary content of the enterprise symbolic architecture. The production environment, construction, advertising, product packaging and product design of the enterprise are the main contents of the corporate symbol culture. Reflecting the spirit of the company with the symbolic form that can be seen and felt. Therefore, the company has designed the product packaging, the clothing brand, the company's advertising and the corporate work environment, re-established relevant standards and improved the overall mental outlook of the company.

Intensive High-Efficiency Green Mining Technology

Focusing on the comprehensive utilization of mineral resources, energy conservation and emission reduction, and developing circular economy as the development goals, we will explore the "five-in-one" intensive and efficient green mining model, fundamentally transform the development mode and economic growth mode, and truly realize the rational development and utilization of resources. Coordinated development with environmental protection. The main features and innovations are reflected in:

Change concepts, innovate mining methods, and achieve comprehensive utilization of resources. Innovate filling mining methods to increase resource extraction rates. According to the particularity of the mine environment, rock conditions and geological conditions, boldly innovate the mining method, completely realize the waste rock without pits, reduce the cost of mining transportation and ground pressure treatment, reduce the loss rate and depletion rate of ore, and improve the mineral resources. Mining rate. Strengthen the utilization of tailings waste rock to realize waste resource utilization. In response to the problem of solid waste destroying the ecological environment, causing environmental pollution and waste of resources, the mine has built a underground mining filling system to strengthen the utilization of tailings waste rock, realizing waste resource utilization, utilization of solid waste such as tailings and waste rock. The level has reached the advanced level of the industry.

Take measures to strengthen emissions control and achieve energy conservation and emission reduction. Strengthen emission control and promote cleaner production. First, according to the production conditions, equipment layout and other on-site conditions, three sets of high-efficiency bag dust removal systems were built, and the dust generated in the belt transportation, crushing and screening of the mine was collected separately, and dust or dust was removed by airtight or suction hood. The equipment purification method is accompanied by spraying atomized water to carry out dust reduction treatment; the second is to build a large tailings concentration tank and circulating water pump pool. The motor control uses inverters to meet the requirements of high concentration of tailings, and the ore recovery wastewater is recovered. Delivered to the production process for recycling to reduce wastewater discharge. The third is to build and install an on-line monitoring facility at the tailings pond wastewater discharge port to conduct real-time online monitoring of emission factors and emissions, and strengthen pollutant emission control management.

Develop water source heat pump technology to solve high-burning coal pollution problems. Cooperated with Shandong Keling Air Conditioning Equipment Co., Ltd., successfully developed water source heat pump technology, using solar energy resources contained in the earth's water body as a source of cold and heat, and energy-converting air-conditioning, is a clean renewable energy technology, without any pollution, and Water and environmental benefits are obvious. The promotion and implementation of this technology solves the problem of high-burning coal and high pollution of enterprises; actively promotes energy-saving lighting and reduces power consumption. Actively research and develop and promote downhole lighting electrodeless lamps, reduce production costs and save energy.

Turn waste into treasure, vigorously develop circular economy, and achieve cost reduction and efficiency increase. Scientific use of equipment waste heat to reduce power consumption. Fuquan Mining fully exploits its potential. For the production wind, the problem of heat loss caused by the centralized air supply from the surface air compressor station is adopted. The air heater cooling is used to carry out the auxiliary warming scheme of the auxiliary well. Through a series of technical transformations, it has been successfully realized. The residual heat of the air compressor meets the preheating requirements of the secondary wellhead in winter, and realizes the zero energy consumption of the winter auxiliary wellhead preheating and wellhead room heating, reducing the manpower input and construction cost; constructing a water resources closed loop system to reduce water waste. For the comprehensive utilization of water resources, the first use of underground production of dewatering water sources in iron mines, extraction of mine water into water containers, the use of water container pressure protection function, directly to the water source heat pump system, to achieve water recycling, It not only can meet the industrial water selection of mines, but also meets the heating, refrigeration and employee bathing water of industrial plants, office areas and some houses; independently develops sewage treatment equipment to realize sewage reuse. Fuquan Mining uses the biochemical treatment technology integrated domestic sewage treatment equipment developed by Sinosteel Maanshan Mine Research Institute to realize domestic sewage reprocessing. After the system is put into operation, it not only protects the environment, but also achieves the goal of wastewater reuse.

Science and technology pioneers, promote technological innovation, and promote enterprise transformation and upgrading. First, Fuquan Mining has always attached importance to the important role of technological innovation in production and environmental protection. In recent years, the proportion of scientific and technological innovation funds in mining output value is much higher than the national standard of 1%, which is maintained at around 3%. Second, continuous improvement and optimization of technological processes. Eliminate outdated processes and production capacity, and production technology ranks at the advanced level of similar mines in China. The high investment in scientific research has brought in many scientific and technological innovations, bringing more benefits to the company. Fuquan Mining has achieved more than 10 technological innovations in mine safety, efficient mining, comprehensive utilization of wastewater, dust control, comprehensive utilization of tailings, and waste rock filling. The application of innovative technological achievements has played an active role in promoting green mine construction.

Establish a system to strengthen scientific management and maintain sustainable development. Fuquan Mining combines production practice. First, in accordance with the relevant provisions of the Green Mines Convention, with the comprehensive utilization of resources and safe production management as the main content, the green mine construction plan is formulated and implemented. Second, according to the plan, nine major management methods, such as mineral resources development and utilization, environmental protection, land reclamation and safe production, have been established, involving more than 60 rules and regulations, detailing the work flow, management and scope of duties of the management work. The assessment method has made the company's work more standardized, scientific and efficient. The establishment and improvement of these rules and regulations has initially formed the modern management system of Sinosteel Fuquan Mining and laid a green mine with low carbon environmental protection and modernization. The cornerstone of efficient and safe operation. The third is to improve the organization of the mining enterprise and implement the target responsibility system, establish a standardized mine management system, clarify the management process and division of responsibilities of the management system, and clarify the duties and regulations and duties of each department and each management position. As well as safety responsibilities, constantly improve the on-site management work mechanism, improve the management level, fully construct and improve the "three-standard integration" comprehensive management system, and establish a continuous promotion mechanism that combines internal audit, management review and external certification.

ROH Safety Production Management System

Applying the modern safety management methods, internal marketization and human resources incentives and other theories and methods, adhere to the management concept of "prevention first" and "continuous improvement", establish and realize the metallurgical mine RHO (risk grading management and control system, hidden danger investigation and management system) Safety management system based on internal market-oriented value incentives. The main features and innovations are reflected in:

Carry out in-depth risk pre-control and pay close attention to accident prevention. Fuquan Mining deeply understands the connotation of "double system" construction of risk grading management and hidden danger investigation and management, accurately grasps the connotation and relationship of risks, hidden dangers and accidents, moves forward to prevent accidents, comprehensively investigates safety risks, and strengthens risk management and control. Use a variety of risk analysis methods (FEMA method, LEC method, etc.) to comprehensively and systematically identify possible accidents, disasters, equipment failures, etc., and pre-assess the possible impacts, giving each process and operation point(Ren,2011). The risks and ratings of the positions, the planning of manpower, material resources, financial resources and technical methods required for risk management and control, and the continuous improvement of methods for the management of hidden dangers, through clear responsibilities, perfect systems, and sound management, to solve the problem of changing the current hidden dangers. Do not go deep, the governance is not thorough, and repeatedly investigate and repeat the problems, and effectively improve the effectiveness of the investigation and management of hidden dangers. On the basis of strictly controlling risks and strengthening the investigation and management of hidden dangers, we will strengthen the pre-disposal of accidents and construct three dangerous and large accident prevention barriers that form risk investigation and control, hidden dangers, and early emergency disposal.

Value quantification stimulates enthusiasm, and safety management is deeply rooted in people Fuquan Mining uses internal marketization theory for safety management, decomposes and refines safety objectives, and quantifies the value of each potential hazard service through internal marketization means. The safety hazards discovered and checked by employees can be verified by the inspectors. Convert to a safe salary. When the safety management personnel enters the underground production site for inspection, the underground behavior of the employees and the safety hazards that may cause accidents at the production site shall be investigated, and the "Safety Service Fees for Production Process" shall be filled in time according to the "Safety Production Standards for Employees". The time, place, hidden dangers, rectification measures, the amount of rewards and punishments, and the responsible person are clearly defined. The difference between the "Production Process Safety Service Fee Schedule" and the previous ticket is mainly in the reward and punishment amount column. The previous ticket will only record the fine amount, and the new checklist will stimulate the enthusiasm of employees to find and find hidden dangers. The penalty amount will be displayed. If it is in a safe state, the inspector will write the bonus amount according to the standard and display it as part of the employee's safe salary. The submitted safety service fee schedule will be compared with the reward and punishment system, and the amount of rewards and punishments for the employees on the day will be calculated. The relevant personnel will fill out the safety and quality standardization class clearing schedule, and finally calculate the safety wages of the day and publicize them on the network. You can check your own daily security salary on the network system. Encourage employees to continuously discover the risks and hidden dangers in their jobs and work in real time, so that "everyone is a safety officer", making safe production a daily life habit that is beneficial to their own value.

Implement the strategy of developing science and technology and build an intrinsically safe mine. Based on information technology, it integrates multiple systems monitoring, monitoring, sound and light alarms such as mining, selection, excavation, machine, transportation, transmission, loading and unloading, and forms a comprehensive safety information management platform with graphics, data, voice and images. . The system is aimed at more than 20 production systems, including main and auxiliary wells, mineral processing, power supply, drainage, ventilation and pressure, etc., effectively connecting the "six systems" such as monitoring and monitoring, personnel positioning, self-rescue, water supply rescue, communication and emergency hedging. It can meet the requirements of "three minutes to inform all personnel in the underground", and is audible, visual and controllable. Once a dangerous situation occurs, you can accurately control the position of personnel and important equipment, get in touch in time, and implement rescue scientifically. The system of power supply, distribution, ventilation, drainage, metering, filling, mineral processing and other systems realizes unattended and remote control, which realizes intrinsic safety, which not only reduces the number of production workers, but also ensures safe production and improves work efficiency. Cooperate with scientific research institutes to implement deformation monitoring system for large-area empty areas, broken roofs and prone to accidental goafs, effectively preventing the occurrence of safety accidents.

Efficient Production of Intelligent Control System

With the goal of industrialization and information integration, with the principle of mechanization substitution, automation and reduction of people, guided by system integration theory and management process theory, the integration of mining, mining, transportation, mineral processing, filling and other production links will be integrated and constructed. An integrated and efficient production management system platform(Yu,2005). The main features and innovations are reflected in:

Top-level planning and design, build an integrated platform for efficient production management and control. In order to carry out overall effective management and control of all kinds of production equipment, production data and operation management of mines, realize all kinds of information integration and standardization of metallurgical underground mines, integration of operation and management, optimization of economic operation, maximization of resource utilization, and interaction of business applications. And decision-making support intelligence, providing strong, reliable, cost-effective support for mining production operations. The intelligent management and control system of metallurgical underground mine production realizes unified management, storage and maintenance of public models, parameters, data and graphics on the basis of overall planning, layered design and unified management.

Based on the new generation of information technology, we will create a one-stop, one-button intelligent mine. Automated, digital, intelligent, unmanned mine construction is a strategic step to achieve the upgrading of the mining industry, representing the direction of today's mine development. Mining combines production practice. First, in accordance with the relevant provisions of the National Informationization Development Strategy and Planning Thoughts of the Ministry of Information Industry, the development and planning of intelligent mine construction is based on the integration of information resources and unmanned production management. Second, according to the requirements of construction planning, actively promote the use of new generation of Internet of Things, big data and other technologies to create a one-stop mine, one-button intelligent mine application platform, through a site to control all production equipment in the entire production process, dynamically master each Information on production processes and links. Start and stop the production equipment in the beneficiation process with one button. Start the production equipment by clicking the "Start" button from the grinding process. The system has interlocking control, fault diagnosis alarm, intelligent selection, priority start and other functions.

CHINA'S SUSTAINABLE DEVELOPMENT OF MINERAL RESOURCES STRATEGIC COUNTERMEASURES

The sustainable development of mineral resources needs to focus on achieving a balance between population growth, environmental conservation and social and economic coordination, and achieving a virtuous circle of resources, environment and social and economic development.

Strengthen Exploration, Increase Capital Investment, and pay Equal Attention to Both Open Source and Expenditure

According to the rich characteristics of China's mineral associated ore and poor ore, strengthen exploration, fully develop lean ore, lean ore, rich and mining, comprehensive survey and evaluation, development and utilization of associated mines, carry out some effective supplementary exploration work, further verify resource reserves, and reserve Low-poor, low-cost mining areas with low efficiency and low efficiency will be shut down; increase capital investment, integrate and utilize domestic and foreign resources, and learn from each other's strengths; but the most important thing is to optimize geological survey work and ensure geological prospecting with the increase of capital investment. There are new breakthroughs, and a number of reserve-type mineral bases will be explored to improve resource reserves. This is the most critical part of open source. Throttling, first of all, must be aware of the need for throttling, and secondly, improve the efficiency of resource use and focus on recycling. Actively develop lean ore, symbiotic ore, associated minerals, should be rich and poor, and open source. It is necessary to comprehensively evaluate and evaluate the efficiency and quality of mining. It is necessary to increase capital investment, optimize geological survey work, and find new mineral bases. Be good at using domestic and foreign resources to improve the shortage of mineral resources. The reserve of mineral resources should be done as soon as possible. It is necessary to pay attention to the development of circular economy and improve the comprehensive utilization efficiency of resources. Focus on recycling. Actively open source and reduce expenditure, and take the road of sustainable development in China.

Demand-Oriented, Global Configuration

China's resource supply strategy should shift from the "self-sufficiency" type in the past to the domestic and global allocation, and maximize the benefits of resources. On the one hand, we diversify, multi-channel and multi-channel use of foreign mineral resources and actively carry out international trade. On the other hand, according to comparative advantages, the trade structure of mineral products is adjusted in a planned and step-by-step manner: importing mineral products that are in short supply in the country and necessary for the national economy, controlling and adjusting the export volume of superior minerals, and consolidating its position in the international market, For maximum economic benefit.

Actively Implement the "Going out" Strategy and the "Introduction" Strategy

On the one hand, we explored the successful way of "going out" and used our political, economic, and diplomatic influences to establish a production and supply base for mineral resources abroad. On the other hand, the country's foreign economic assistance and the "going out" strategy are organically combined to establish a stable, economical, and on the premise of enhancing the competitiveness of mining companies in the international market and China's influence on global mineral resources. A diverse mineral supply system. Accelerate the development of China's opening up to the outside world and strengthen exchanges with the world. However, at present, China is still not mature in introducing foreign capital, and there are still some problems in introducing and developing mineral resources. Therefore, we should further improve the investment environment, further enrich our benefits and attract them to invest. Actively encourage foreign businessmen to invest in China and develop new minerals for foreign cooperation, and increase opportunities for cooperation. Give them a certain amount of excellent energy. Accelerate the development of China's economy and promote the sustainable development of the economy. At the same time, increase the "introduction", promote the use of foreign capital for commercial mineral exploration; strengthen cooperation with neighboring countries' mineral resources, especially energy, fully open offshore mineral exploration and development, and further promote the upgrading and structure of China's minerals. Adjustment.

Rely on Technological Progress and Improve the Comprehensive Utilization Rate of Resources

Although there are many mineral sources in China, it is more poor ore. The choice of mineral resources provides a platform for talents to develop, which not only expands the exchanges between industries, but also improves the level of employment of relevant personnel. A subsidy. China's mineral source multi-component ore, difficult to mine, rich ore, and the fundamental way of mining development and utilization is to improve the level of mineral processing, relying on scientific and technological progress, fully develop and utilize lean ore, and fully improve the comprehensive utilization of mineral resources; rely on Scientific and technological innovation, improve the quality of geological and environmental work, and implement important ways to protect the environment, conserve resources, develop clean production, ensure the recycling and harmlessness of waste, and promote the sustainable development of mining areas.

Improve Policies and Regulations and Improve the Market System to Provide Environmental and Safety Guarantees

China is a big energy consumer, and domestic energy supply is in short supply, especially oil and energy. And the gap is getting bigger and bigger. In order to maintain the stability of China's petroleum energy use, we should start from China's national conditions and establish a mineral resource reserve system suitable for China. For scarce resources, we should actively implement a protective development strategy. Promote the healthy and sustainable development of the national economy. Establish a stable mining order, effectively stop the phenomenon of indiscriminate mining, provide legal protection for the ecological environment protection and comprehensive utilization of resources in the mining area, and conscientiously implement the "Ministry of Mineral Resources Law of the People's Republic of China" on "the implementation of mineral exploration and development by the state" The principle of comprehensive survey, unified planning, rational mining, rational layout and comprehensive utilization. Establishing a sound management system quality and skills can better conserve resources and protect the environment. Improve laws and regulations and improve management systems. Establish a stable mining order, strengthen supervision over the phenomenon of indiscriminate mining, and do well in advance control. We will curb the occurrence of indiscriminate excavation, strengthen the protection of the ecological environment and the comprehensive utilization of resources, and provide legal protection for it. The relevant provisions of the Mineral Resources Law should be implemented, and the survey work should be actively carried out, unified planning, and rational development and utilization. Improve the efficiency of resource utilization, and use strong institutions to regulate behaviors and promote sustainable economic development.

CONCLUSION

In summary, the lack of resources and energy is an unchangeable fact, and the large demand for mineral resources caused by the improvement of people's living standards has become an unavoidable trend. In order to meet people's needs for mineral resources and sustainable development of resources, the management concept of the mineral industry also needs a certain degree of innovation. Therefore, deepening the ideological construction of the concept of sustainable development, improving mining efficiency and seeking new energy alternatives have become the current trend. Only by continuously innovating and reforming with the development of society can we promote the sustainable development of mines without unduly destroying the ecological environment. The development of mineral resources is a long-term process. We should formulate development strategies suitable for ourselves according to China's national conditions. We should actively do both open source and cost reduction, strive to attract foreign investment, and rely on the advancement of science and technology and the improvement of the system to promote sustainable economic development.

REFERENCES

Babie, P. (2013). Sovereignty as governance: an organising theme for australian property law. Social Science Electronic Publishing, 36(3).

Carayannis, E. G., & Alexey, E. (2017). Sustainable development of the russian arctic zone energy shelf: The role of the quintuple innovation helix model. *Journal of the Knowledge Economy*, *8*(2), 456–470. doi:10.100713132-017-0478-9

Drielsma, J. A., Russellvaccari, A. J., Drnek, T., Brady, T., Weihed, P., Mistry, M., & Simbor, L. P. (2016). Mineral resources in life cycle impact assessment—Defining the path forward. *The International Journal of Life Cycle Assessment*, *21*(1), 85–105. doi:10.100711367-015-0991-7

Gassman, P. W., Williams, J. R., Benson, V. W., Caesar Izaurralde, R., Hauck, L. M., & Jones, C. A. (2005). *Historical development and applications of the epic and apex models.* Center for Agricultural and Rural Development (CARD) Publications.

Halme, M., & Korpela, M. (2014). Responsible innovation toward sustainable development in small and medium-sized enterprises: A resource perspective. *Business Strategy and the Environment*, *23*(8), 547–566. doi:10.1002/bse.1801

Lei, L., Zhou, J., Dong, J., Zhuang, D., Mansaray, L. R., Hu, Z., & ... . (2016). Mineral resources prospecting by synthetic application of tm/etm+, quickbird and hyperion data in the hatu area, west junggar, xinjiang, china. *Scientific Reports*, *6*(6), 21851. PMID:26911195

Liu, C., & Han, G. (2008). The assessment of innovation capacity of high-tech industries with the view of sustainable development: an empirical study in china. In *IEEE International Conference on Industrial Engineering & Engineering Management*. IEEE.

Liu, L., Qiao, Y., & Hao, Z. (2016). Black carbon concentration and isotopic composition of surface sand from deserts and dune fields in northern china. *Palaeogeography, Palaeoclimatology, Palaeoecology*, *445*, 1–7. doi:10.1016/j.palaeo.2015.12.026

Moosavi, S. C. (2015). Minnesota mineral education workshop professional development for teachers in the area of mineral resource education. *Journal of Orthopaedic Research*, *33*(9), 1382–1390. PMID:25808101

Niekerk, V., & Heleen. (2015). Unlocking mineral-tenure security: the application procedures under the mineral and petroleum resources development act 28 of 2002. *Journal of African Earth Sciences, 12*(s 1–2), 283-295.

Ren, Xiao-yu, Yang, & Xi-huai. (2011). A Study of Mineral Resources Mining Model from the Perspective of Sustainable Development. *Proceedings of the 2011 Chinese Control and Decision Conference(CCDC).*

Rossi, M. E., & Deutsch, C. V. (2012). *Mineral Resource Estimation*. Academic Press.

Sharfman, M. P., Meo, M., & Ellington, R. T. (2000). Regulation, business, and sustainable development: The antecedents of environmentally conscious technological innovation. *The American Behavioral Scientist*, *44*(2), 277–302. doi:10.1177/00027640021956215

Sonderegger, T., Pfister, S., & Hellweg, S. (2015). Criticality of water – aligning water and mineral resources assessment. *Environmental Science & Technology*, *49*(20), 12315–12323. doi:10.1021/acs.est.5b02982 PMID:26392153

Vollenbroek, F. A. (2002). Sustainable development and the challenge of innovation. *Journal of Cleaner Production*, *10*(3), 215–223. doi:10.1016/S0959-6526(01)00048-8

Wang, Z., Kusky, T. M., & Capitanio, F. A. (2016). Lithosphere thinning induced by slab penetration into a hydrous mantle transition zone. *Geophysical Research Letters*, *43*(22), 11,567–11,577. doi:10.1002/2016GL071186

Yu, J., Yao, S., Chen, R., Zhu, K., & Yu, L. (2005). A quantitative integrated evaluation of sustainable development of mineral resources of a mining city: A case study of huangshi, eastern china. *Resources Policy*, *30*(1), 7–19. doi:10.1016/j.resourpol.2004.08.006

Zhang, S., Xiao, K., Zhu, Y., & Cui, N. (2017). A prediction model for important mineral resources in china. *Ore Geology Reviews*.

Chapter 6
Utilization of Agro Waste:
An Approach for the Administration of Healthy Environment and Public Domain

Savita Budhwar
Central University of Haryana, India

Manali Chakraborty
Central University of Haryana, India

ABSTRACT

Agro-wastes are one of the major sources for nutritional and therapeutic purposes along with other beneficial properties according to some ongoing studies. Yet they are not being utilized properly; instead, these wastes are being discarded in open air, causing pollution. Conservation of such natural resources is a healthy approach, and it's economic and eco-friendly. too. Novel formulated food products from these agro-wastes can be utilized as an unconventional source of nutrients and a therapeutic weapon for diseases such as diabetes, hypertension, etc. along with normal human consumption. Utilization of such agro-wastes as food products will also be useful to combat malnourished state leading to rural development and maintenance of food security in cheap prices for a healthy seeks. It also shows a possible role as antibiotics, antioxidants, anti-inflammatory agents, adsorbents, enzyme extracting sources, insulators, and food additives along with its contribution in solid state fermentation, beneficial in dairy industries.

DOI: 10.4018/978-1-7998-1966-0.ch006

INTRODUCTION

Recognition of the by products resulting from the industrial or societal process is very much crucial for the waste management instead of disposing them into air or water, burying in land or burn them. They may be a very good source of nutrients. Sole composition of these agro wastes may not be the appropriate balance of nutrients for the other utilizations such as, value added food production; combined with one another can be low cost sources of health benefits. The most economical method to dispose rice husk is considered to burn it to ashes. This husk ash from the milling industry as well as the field is causing environmental pollution leading to major difficulty. Major economic way to recycle these wastes is the formation of value added products (Anuj kumar et al. 2015). This husk ash contains silica predominantly along with other elements such as potassium, magnesium, calcium. Dumped wastes in open environment causes water as well as land pollution becoming one of the causes of global climatic changes in twenty first century (Shi et al. 2010). Gathering attention to dispose in innovative ways and utilizing these agro wastes is really a matter of concern (Zhang et al 2015).

Reducing agro losses during industrial processing and to make use of by-products appropriately for human feed is now a major global alarm. Before utilizing such wasted parts their nutrient as well as anti-nutrient breakdown is very much needed to understand to check the consumable properties along with toxicities, if any.

Almost 70% yield of rice as the major one is recorded and rests are unconsumed part, utilized as animal feed. These wastes consist of rice husk, rice bran and rice germ. During polish of rice 1-17% broken rice is also generated which is consist of broken kernels. Almost 2.95 million tons de-oiled rice bran production has been recorded just for animal fed. Mung bean bran, also called chunni in India, consist of broken parts of bean, is rich in protein. 50% of the buffaloes feed is consist of bran of mung maintaining proper nutrition in cattle fed (Krishna et al. 2002).Chickpea by-products (bran, straw and husk) are highly nutritive and a large amount of it is consumed by animals. Almost 10-11% bran is produced during processing (Kanatt et al. 2011).

Global population and its subsequent increment is one of the most critical challenges for the future. Additional food demand, malnourished condition is going to create a terrible concern in the coming timeline of human kind until and unless maintenance of post-harvest losses and food security is considered from now only. It is expected that population mark will rise up to 9.1 billion people by the year 2050 and almost 70% extra food requirement will be needed to avoid hunger issues (Godfray et al. 2010, Hodges et al. 2011, Parfitt et al. 2010). Though developing countries already started to improve their agricultural condition and productions, awareness regarding post-harvest loss (PHL) is still not sufficient. Less than 5%

research funding has been issued in this context in previous few years (Kitinoja et al. 2011). Balance of this food demand along with the waste management to control environmental pollution is undoubtedly a critical research priority. This environment friendly approach not only might be a rich source of nutrients as food substances but also can be transformed into marketable valuable products emerging low cost budget economical independency for the below poverty lined population.

By product of plant food processing represent a major disposal problem for the industry concern, but they are also promising source of compounds which may be used because of their favorable technological and nutritional properties. On account of almost stagnant pulse production, increased population and environmental health issues related to burning of these waste, it has become necessary to improve the processes to utilize the milled, broken, husk part and prepare supplementing and mixed by product foods to meet the current dietary demands of human beings.

Approximately 20-30% of by products have not been utilized. Even though utilized but their full potential as a source of nutrient has not been exploited. Therefore they are underutilized and degraded in environment.

In current scenario, we need methods and process to save unused resources for betterment of society. Value added products developed from waste generated from by- products of Bengal gram, wheat bran, rice bran, broken rice can have potentials benefits for weight loss therapy, diabetes, hypocholesterolemic patients and malnourished children. Scavenging their nutrient content for the development of the food products might nullify the negative impact of the agro waste disposal over environment. Hence in the proposed chapter an attempt will be made to discuss about various technologies regarding value addition of the agro wastes, viz., by-product of cereals, pulses etc to maintain nutrition security instead of degraded into the environment.

BACKGROUND

Recognition of the by products resulting from the industrial or societal process is very much crucial for the treatment of the wastes instead of disposing them into air or water, burying in land or burn them. They may be a very good source of nutrients. Sole composition of these agro wastes may not be the appropriate balance of nutrients to utilize for other purposes such as, value added food production; combined with one another can be low cost sources of health benefits. The rice husk ash contains silica predominantly along with other elements such as potassium, magnesium, calcium. Dumped wastes in open environment causes water as well as land pollution becoming one of the causes of global climatic changes in twenty first century (shi et al. 2010). Gathering attention to dispose in innovative ways and utilizing these agro wastes is really a matter of concern (Zhang et al 2015).

Absence of appropriate primary processing technologies to prepare ready-to-use or ready-to-cook (RTC) products and also secondary as well as tertiary processing to prepare ready-to-eat value added products have been the major limiting factors for their diversified food uses and better economic status (Malleshi et al. 2014). Antioxidant rich rice husk contains polyphenols that shows high activity against scavengers of singlet oxygen and to inhibit hydrogen peroxide activity to protect DNA damage (Jeon et al. 2006). They also contain many phytochemicals viz., alpha, beta, gamma, delta-tocopherol, tocotrienols (vitamin E) and gamma-oryzanol. Not only is the husk part, rice bran layer also rich in antioxidant contents. Besides rice paddy, wheat is also a primary cereal crop, mainly used for human consumption and domestic animals feed. 90% bran is utilized for this purpose and rest 10% for food products in spite of having rich nutrients.

Especially the outer layer of wheat grain contains high level of antioxidants, vitamins (Vit-E, B), phenolic compounds, sterols, minerals, insoluble dietary fibers. Its nutritional composition also includes bioactive compounds viz. alkylrescorcinols, ferulic acid, flavonoids, carotenoids, lignans, also called phytochemicals (De Brier et al. 2014). Phenolic acids in bran part remain in bound and free, both states with approximately 99% of ferulic acid in insoluble and bound situation (Lu et al. 2014).

Pulses like Bengal gram (Chickpea), mung bean etc are also in parallel competition to provide food security through utilization of its waste parts. Other highly nutrient fraction of bean is its hull, straw, meals, which are being used mainly for livestock consumption, could be utilized proficiently as human fed by managing anti-nutrient properties. Bengal gram is considered to be the fourth largely produced pulse crop in the World. A novel antioxidant present in husk is methoxy pectin in low scale, related

Figure 1. Percentage of wheat grain parts (Surget et al. 2005)

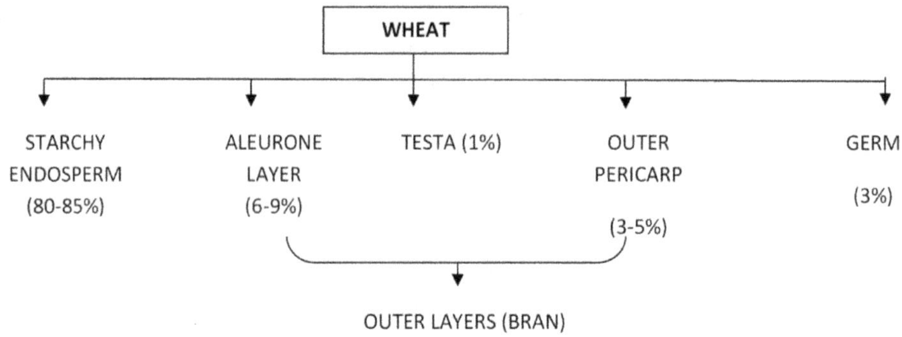

to galacturonic acid content (Vania et al. 2010). Utilization of these chemical and physical properties of agro wastes to form foodstuffs is the key purpose of this study.

Global population and its subsequent increment is one of the most critical challenges for the future. Additional food demand, malnourished condition is going to create a terrible concern in the coming timeline of human kind until and unless maintenance of post-harvest losses and food security is considered from now only (Godfray et al. 2010, Hodges et al. 2011, Parfitt et al. 2010). Though developing countries already started to improve their agricultural condition and productions, awareness regarding post-harvest loss (PHL) is still not sufficient (Kitinoja et al. 2011).

The most economical method to dispose rice husk is considered to burn it to ashes. This husk ash from the milling industry as well as the farmers is causing environmental pollution leading to major difficulty. Major economic reason to recycle these wastes is the formation of value added products (Anuj kumar et al. 2015).

Reducing agro losses during industrial processing and to make use of by-products appropriately for human feed is now a major global concern. Before utilizing such wasted parts their nutrient as well as anti-nutrient breakdown is very much needed to understand their consumable properties along with toxicities, if any. According to GAIN (Global Agricultural Information Network) report (IN7031), marketing year 2017/18 rice production estimation is at 106 million metric tons (MMT) from 44.5 million hectares. Almost 70% yield of rice as the major one is recorded and rests are unconsumed part, utilized as animal feed. These wastes consist of rice husk, rice bran and rice germ. During polish of rice 1-17% broken rice is also generated which is consist of broken kernels.

During harvesting, storage and milling though there is a mild nutrient loss, which is considerable. Due to polishing process, almost 29% of protein content is lost along with 79% of fat and 67% of iron (Aadil et al. 2011). Almost 2.95 million

Figure 2. Quantity of by-products from rice (Cereal) during processing

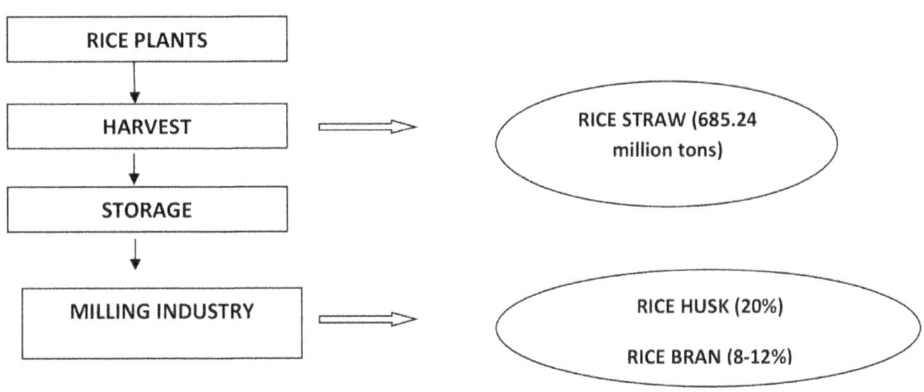

tons de-oiled rice bran production has been recorded just for animal fed. Besides having rich nutrient value, rice by-products also consist of some anti-nutrients viz. lipases, trypsin inhibitor (85%-95%), hemagglutinin-lectin and phytates. It has been observed that there is a rapid increment in the free fatty acid level within hours and level reaches upto 7%-8% within 24 hours, followed by 5% increment per day.

Moisture, temperature, pH, time, water activity affects lipase activity on a large scale. Occurrence of phytic acid in addition affects the solubility, functionality and digestibility of proteins and carbohydrates (Ramzan et al. 2000).Due to presence of oryzanol, rice by-product has cholesterol lowering properties and hypoglycaemic effect (hyperglycemia induced neuropathy) (Siddiqui et al. 2010).Due to presence of these potential components in wheat bran it could be beneficial to lower blood cholesterol level, cardiovascular diseases, type II diabetes, cancer prevention and many others (Javed et al. 2012). In India, approximately 3.70 million ha is grown and the annual estimation is 1.57 million tons (Sharma et al. 2011). By-products are usually being used for animal feeds. Mung bean bran, also called chuni in India, consist of broken parts of bean, is rich in protein. 50% of the buffaloes feed is consist of bran of mung maintaining proper nutrition in cattle fed (Krishna et al. 2002). Chickpea by-products (bran, straw and husk) are highly nutritive and a large amount of it is consumed by animals. Almost 10-11% bran is produced during processing (Kanatt et al. 2011).

MAIN FOCUS OF CHAPTER

Issues, Controversies, Problem

Nutrient rich agricultural wastes that are generated during the milling process or processing of the agricultural contents are matter of concern for nutrient security as well as environmental cleanliness. Dumping processes of these agro wastes generate hazardous constituents that lead to pollution. For an example, burnt rice husks causes difficult problems when it comes about its disposal. A significant part of the produced rice husk is utilized as livestock feed but most of them are being dumped off as waste or the purpose of landfill or burnt to generate power for the mills. This agro waste seems to be of very hard surface, contains high silica along with small bulk density and proved to be a dangerous resource of water and land pollution in open environment (Kalapathy et al. 2000). Now-a-days unavailability of pollution free water is a crucial need for this generation as well as in near future (Shi et al. 2010).

Few agro wastes already created a lot of controversies due to their carcinogenic and bio-accumulative nature after burning them into ecosystem. This seems to cause various health related problems resulting into fatal diseases, such as loss of appetite, fatigue, respiratory problems, skin irritation and even death (Zhang et al. 2015).

Thus proper disposal through innovative technologies of these kinds of agricultural wastes are very much important for their better utilization. Besides having health effecting contents they are also rich source of nutrients and this untapped nutrition might be utilized to combat malnutrition and hunger. More over development of such innovative technologies of such crucial agro wastes may also contribute in the economical status of concerned area.

SOLUTIONS AND RECOMMENDATIONS

Their uses are not only limited up to nutrition security but also they are considered to be great source for other purposes. Various contamination and discharge of different toxic contents from industries in water, soil, and atmosphere are alarming. Regarding the possible treatment of these pollutants, some techniques are really helpful, viz., coagulation, flocculation, sedimentation, biodegradation, chemical adsorption, ion exchange techniques etc. (Foo and Hameed 2009a; Kumar et al. 2009). But their efficiencies are not that promising except the adsorption (Daifullah et al. 2004; Doke and Khan 2013). Activated carbon is being used for such purposes. Based upon some studies, researchers are now showing more interest to generate this activated carbon by utilizing low cost materials, such as, orange peels, sugar cane bagasse, maize etc by chemical activation and heat treatment, also known as carbonization (Dasgupta et al. 2015; Patel 2012; Song et al. 2015). Rice husk also showed promising effects as adsorbent and the most important thing in case of using husk as adsorbent is, it is economically feasible as well as this utilization of husk is also gives a solution to make environment pollution free from disposal issues (Foo and

Hameed 2009c). Extracted silica as sodium silicate from rice husk is seemed to be economical and this make husk an alternative useful source of silica, that is being used for enormous purposes, viz., transistor, diodes, electronic circuits, thermal insulator and so on. Others uses of rice husk as its ash form (Rice husk ash/ RHA) also include concrete for buildings, preservatives to save crops or foods from insect attacks etc.

Approximately 147.2 million metric tons of fiber sources, 709.2 and 673.3 million metric tons of wheat and rice straws are generated respectively worldwide (Belewu and Babalola 2009). These agro by-products contains anti oxidant properties, anti inflammatory activity along with health beneficial macro and micro nutrients. Wastes from plant sources, such as, pomegranate peels, lemon peels, green walnut husks are seemed to be promising based upon recent studies as natural antimicrobials (Adámez et al. 2012; Katalinic et al. 2010). These wastes also can be utilized as a

Figure 3.

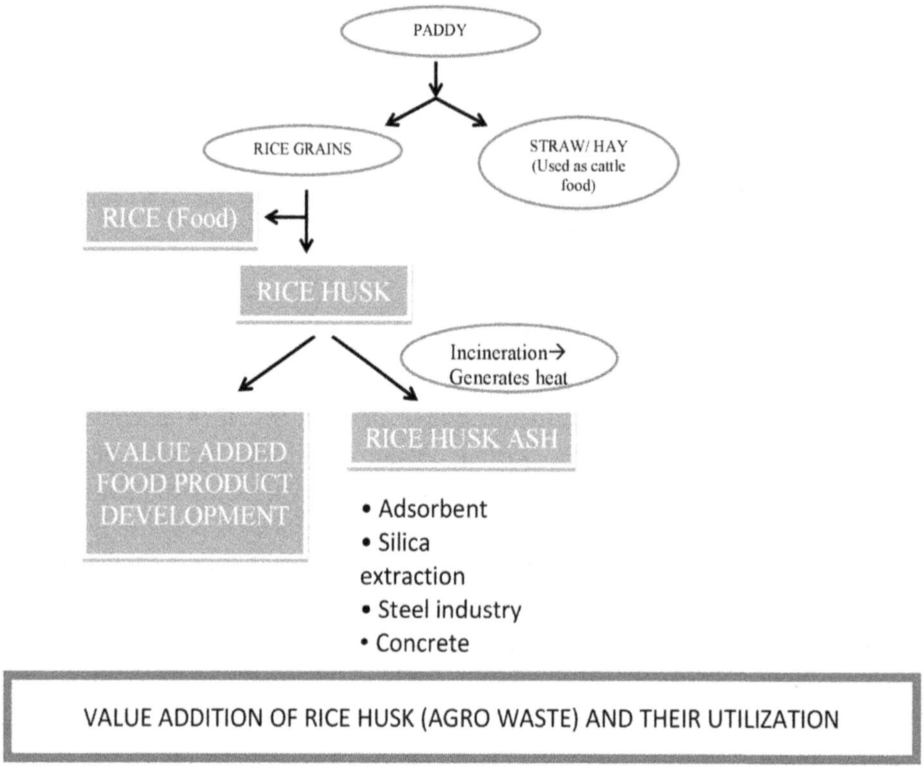

medium for the mushroom cultivation along with the development of value added food stuffs. They are also considered to be an appropriate medium for the growth of microorganism further helping in the production of fermentable sugars that in turn cut the production cost (Nguyen et al. 2010).

Generated agricultural wastes can be characterized based upon their mode of production. They could be field residues or process residues. Wastes that remain after the processing of crop harvesting are generally known to be as field residues (leaves, stalks, seed pods, stems). Whereas, process residues remain present till after the successful alteration of crops into other valuable sources. Generally they are used for animal feed or as soil fertilizer (Zafar 2014).

These agro wastes being appropriate support for the micro organism growth, also contributes in the production of enzymes. They are considered to be low cost source for the value addition.

Several food flavouring agents can be produced by using up these agro wastes. Researchers found out biotechnological techniques to recover ferulic acid from food

Figure 4.

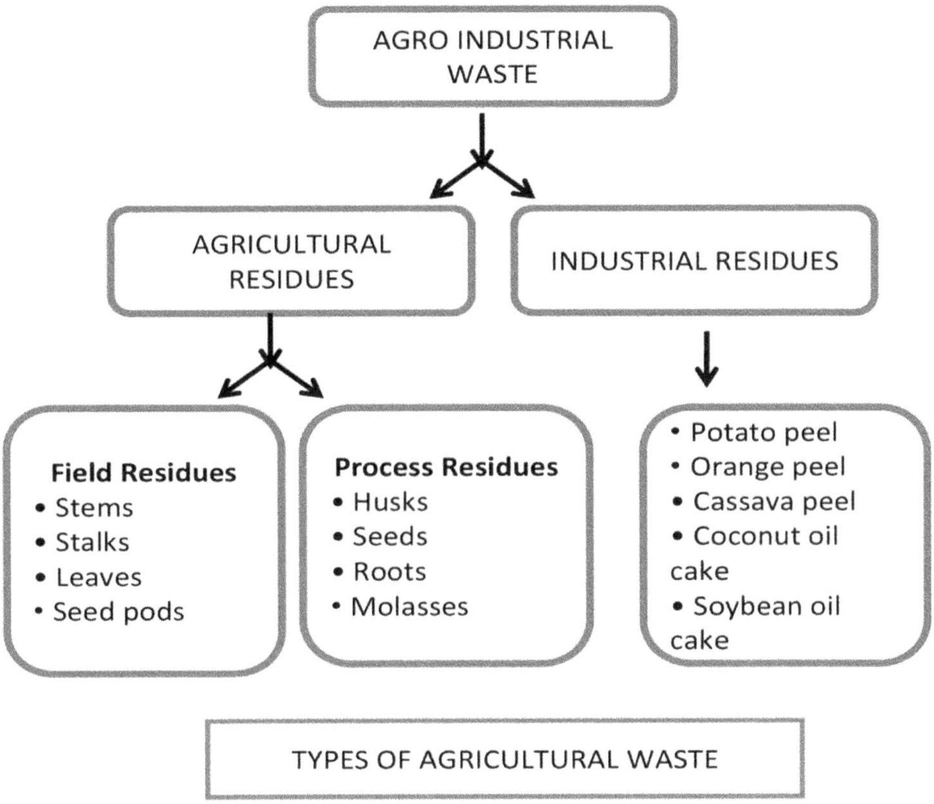

industry by- products and its considered as economical, environmentally sustainable (Ayala- Zavala JF et al, 2009).

Wastes are generated from different phases of food cycle too, such as, agricultural production, processing and distribution. Development of various methods to utilize these wastes is attempted for therapeutic use from time to time. These are generally used for animal feed and fertilizers. But based upon further researches it has been observed that these wastes are way too much rich in nutrients that can be used for the formulation of high valued products (Rudra et al. 2015).

For decades researchers are searching for natural remedy and natural bioactive resources to prevent diseases and to lessen toxic side effects. Agricultural plant wastes are found to be one that can efficiently interact with biomolecules, viz. DNA, and proteins producing desired result and can be further used for designing therapeutic agents (Ajikumar et al. 2008) that is functional food products. Phytochemicals, antioxidants extracted from these sources are not only found to be enhancing positive effect on human health but also they boost up other biological function,

Table 1. Chemical composition of various agricultural wastes

Agro- industrial wastes	Chemical composition (% w/w)						References
	Cellulose	Hemicelluloses	Lignin	Ash (%)	Total solids (%)	Moisture (%)	
Sugarcane bagasse	30.2	56.7	13.4	1.9	91.66	4.8	El- Tayeb etal (2012) and Nigam et al. (2009)
Rice straw	39.2	23.5	36.1	12.4	98.62	6.58	El- Tayeb etal (2012)
Corn stalks	61.2	19.3	6.9	10.8	97.78	6.40	El- Tayeb etal (2012)
Sawdust	45.1	28.1	24.2	1.2	98.54	1.12	El- Tayeb etal (2012) and Martin et al (2012)
Sugar beet waste	26.3	18.5	2.5	4.8	87.5	12.4	El- Tayeb etal (2012)
Barley straw	33.8	21.9	13.8	11	---	---	Nigam et al. (2009)
Cotton stalks	58.5	14.4	21.5	9.98	---	7.45	Nigam et al. (2009)
Oat straw	39.4	27.1	17.5	8	---	---	Martin et al (2012)
Soya stalks	34.5	24.8	19.8	10.39	---	11.84	Motte et al (2013)
Sunflower stalks	42.1	29.7	13.4	11.17	---	---	Motte et al (2013)
Wheat straw	32.9	24.0	8.9	6.7	95.6	7	Nigam et al. (2009) and Martin et al (2012)

such as immunity, digestion, growth and development. This new concept of food products also grabbed a huge attention in food market due to the increased consumer interest for "healthy" food. Instead for chemically synthesized these functional foods can be used to treat malnutrition and nutrient deficiency (Joana Gil-Chávez et al. 2013). Various wastes from vegetables, like, stems, seeds, shells, bran and residue remaining can be used for value added product formulation in food processing as well as pharmaceutical preparation (Baiano 2014). Tomato by- products are found to be a great source of bioactive compounds like phyto chemicals, viz., tocopherols, carotenes, terpenes and polyphenols. These can also be a great replacement of

Table 2. Chemical composition of fruit industrial wastes

Fruit Industrial wastes	Chemical composition (% w/w)								References
	Cellulose	Hemi-celluloses	Lignin	Ash	Total solids	Moisture	Total carbon	Total nitro-gen	
Potato peel waste	2.2%	---	----	7.7%	----	9.89	1.3%	----	Weshahy and Rao (2012)
Orange peel	9.21%	10.5%	0.84%	3.5%	----	11.86	----	----	Rivas et al (2008)
Coffee skin	23.77 (g/ 100g)	16.68 (g/ 100g)	28.58 (g/ 100g)	5.36 (g/100g)	----	----	C/N 14.41	----	Lina et al (2014)
Pineapple peel	18.11	----	1.37	---	93.6	91	40.8	0.99	Paepatung et al (2009)

Figure 5. Agricultural wastes (orange peel, rice bran, wheat bran)

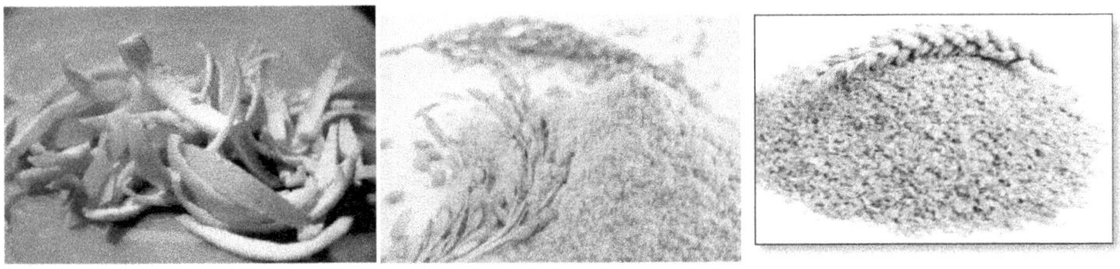

chemical additives to enhance shelf life of foods (Kalogeropoulos et al. 2012). Phenolic compounds, carotenoids, vitamin C and dietary fibre are extraxted from mango peel that can contribute in lowering health risk in cancer, cataracts, alzheimer's disease and Parkinson diseases (Ayala- Zavala et al. 2012).

FUTURE RESEARCH DIRECTIONS

Various studies are going on to maintain nutrition as well as food security. Reporting different novel techniques to extract and quantify the bioactive molecules from the agricultural wastes along with agro- industrial residues is one of them. Availability of these highly promising technologies is providing a significant opportunity to quantify different ranges of metabolites with accuracy. Based upon the recorded range the by- products can be utilized more specifically in various dimensions. Though most of the extraction methods are able to give the accurate, supercritical fluid

Table 3. Enzyme production by utilizing agro wastes

Substrate	Enzymes	Microorganisms	Source
Papaya waste	Alpha- amylase	A. *niger*	Sharanappa et al (2011)
Groundnut oil cake	Lipase	C. *rugosa*	Rekha et al (2012)
Wheat bran and orange peel	Pectin methyl esterase	P. *notatum*	Gayen and Ghosh (2011)
Linseed oil cake	Lipase	P. *aeruginosa*	Dharmendra (2012)
Orange peel	Alpha- amylase	A. *niger*	Sindiri et al (2013)
Coconut oil cake	Alpha- amylase	A. oryza	Ramachandran et al (2004)
Rice bran	Alpha- amylase	*Bacillus sp.*	Sodhi et al (2005)
Corn bran	Alpha- amylase	*Bacillus sp.*	Sodhi et al (2005)
Rice bran, wheat bran, black fram bran and soybean	Alpha- amylase	A. *niger*	Akpan et al (1999)
Fruit peel waste	Invertase	A. *niger*	Mehta and Duhan (2014)

Figure 6.

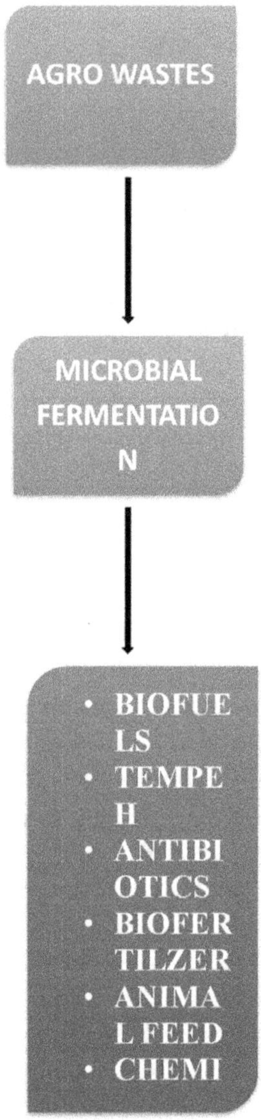

DIFFERENT USAGE OF AGRICULTURAL WASTES BY MICROBIAL FERMENTATION

Table 4. Supercritical technologies for the value addition in agro wastes

Applied technology	Source	Target compounds	Process condition	Reference
Subcritical water hydrolysis	Coconut husk, defatted grape seed, and pressed palm fiber	Fermentable sugars	Pressure 20Mpa Temperature 208-257ºC	Prado et al. 2014
Subcritical ethane extraction	Tomato wastes	Lycopene	Pressure 30Mpa Temperature 60ºC	Nobre at al. 2012b
Supercritical antisolvent extraction	Grape seeds	Antioxidants	Pressure 8-15Mpa Temperature 35-60ºC	Marqués et al. 2013
Pressurized liquid extraction (PLE)	Potato peels	Anthocyanins	Pressure 10Mpa Temperature 80ºC	Casas Cardoso et al. 2013
Pressurized liquid extraction (PLE)	Jabutic- aba skins	Anthocyanins	Pressure 5-10Mpa Temperature 40-120ºC	Santos, Veggi, and Meireles 2012
Supercritical CO2 extraction	Grape seeds	Proanthocyanidin	Pressure 25MPa, 27.5MPa,30MPa Temperature 30ºC,50ºC	Yilmaz, Ozvural, and Vural 2011
Supercritical CO2 extraction	Tomato peel	Lycopene	Pressure 20-40Mpa Temperature 70-90ºC	Machmudah et al. 2012
Supercritical CO2 extraction	Grape seeds	Phenolic compounds and antioxidants	Pressure 15.3- 16.1Mpa Temperature 44-46ºC	Ghafoor, Al-Juhaimi, and Choi 2012
Supercritical CO2 extraction	Grape bagasse	Polyphenol	Pressure 20-35Mpa Temperature 40ºC	Farías-Campomanes, Rostagno, and Meireles 2013
Supercritical CO2 extraction	Jabutic- aba skins	Anthocyanin	Pressure 10-35Mpa Temperature 40-50ºC	Santos et al. 2013
Supercritical CO2 extraction	Blackberry bagasse	Antioxidant compounds	Pressure 15Mpa Ultrasonic power 800W	Reátegui et al. 2014
Supercritical CO2 extraction	Coffee grounds	Oil and diterpenes	Pressure 14-19Mpa Temperature 40-70ºC	Barbosa et al. 2014

continued on following page

Table 4. Continued

Applied technology	Source	Target compounds	Process condition	Reference
Supercritical CO2 extraction	Coffee grounds	Triacylglycerides	Pressure 19Mpa Temperature 40-55°C	Melo et al. 2014
Supercritical CO2 extraction	Passion fruit seeda	Seed oil	Pressure 16Mpa, 26Mpa Ultrasonic power 0W, 160W, 640W	Barrales, Rezende, and Martínez 2015
Supercritical CO2 extraction	Hazelnut, coffee and grape wastes	Triglycerides and polyphenols	Pressure 35-50Mpa Temperature 40-60°C	Manna, Bugnone, and Banchero 2015
Supercritical CO2 hydrolysis	Sugarca- ne bagasse	Fermentable sugars	Pressure 10-25Mpa Temperature 40-80°C	Benazzi et al. 2013
Supercritical CO2 hydrolysis	Peach seeds	Oil and beta-sitosterol	Pressure 16-24Mpa Temperature 35-55°C	Ekinci and Gürü 2014
Subcritical water extraction	Onion skin	Flavonol quercetin	Pressure 9-13.1Mpa Temperature 100-190°C	Ko et al. 2011
Subcritical water extraction	Potato peel	Phenolic compounds	Pressure 6Mpa Temperature 100-240°C	Singh and Mda 2011
Subcritical water extraction	Pomegranate seeds	Phenolic compounds	Pressure 6Mpa Temperature 80-280°C	He et al. 201
Subcritical water extraction	Winery wastes	Antioxidants	Pressure 8Mpa, 11.5Mpa, 15Mpa Temperature 100°C, 120°C, 140°C	Aliakbarian et al. 2012
Supercritical water hydrolysis (continuous system)	Corn residue	Hexoses	Pressure 23-24Mpa Temperature 380°C	Zhao et al. 2012
Subcritical water extraction (batch)	Sugarca- ne bagasse	Reducing sugar	Temperature 200-240°C Reaction time 120s	Zhu et al. 2013

Table 5. Application of various agro by- products as additives in foods

Source material	Extracted compounds	Food	Function	Reference
Grape seed	Antioxidants	Corn chips	Preventing lipid oxidation	Rababah et al. 2011
Grape seed	Polyphenols	Yogurts	Producing functional yogurts	Chouchouli et al. 2013
Grape seed	Antioxidants	Frozen precooked chicken nuggets	Inhibiting lipid oxidation	Cagdas and Kumcuoglu 2015
Wine grape pomace	Antioxidants dietary fibre	Yogurt and salad dressing	Enhancing nutritional value and improving storability	Tseng and Zhao 2013
Wine grape pomace	Bioactive compound	Fortified baked goods	Improving physicochemical, nutritional, and sensory qualities	Walker et al. 2014
Wine grape pomace	Antioxidants	Beef patties	Reducing protein oxidation	Garcia- Lomillo et al. 2016b
Wine grape pomace	Antioxidants	Barbecued beef patties	Reducing polycyclic aromatic hydrocarbons (PAHs) and heterocyclic aromatic amines (HAs)	Garcia- Lomillo et al. 2016b
Wine grape pomace	Antioxidants	Refrigerated and frozen beef patties	Inhibiting lipid oxidation	Garcia- Lomillo et al. 2017
Grape marc	Polyphenols	Fettuccini pasta	Increasing antioxidant activity	Sant'Anna et al. 2014
Orange byproducts	Dietary fibre	Ice cream	A fat replacer	Crizel et al. 2013
Orange byproducts	Dietary fibre	Yogurts	A fat replacer	Yi et al. 2014
Brans	Dietary fibre	Pasta	Producing functional pasta	Kaur et al. 2012
Brans	Dietary fibre	Bread	Improving bread- baking qualities	Cai et al. 2014
Apple, banana or passion fruit wastes	Dietary fibre	Yogurts	Enhancing probiotic validity and fatty acid profile and increasing CLA content	do Espirito Santo et al. 2012a
Passion fruit peel	Dietary fibre	Yogurts	Producing skim yoghurts	do Espirito Santo et al. 2012b
White cauliflower byproducts	Antioxidant compounds	Beef sausage	Improving the nutritional, physicochemical and sensory quality	Abul- Fadl 2012
Old stalks of Asparagus officinalis	Total flavonoids and total phenolic	Restructured chips	Producing restructured asparagus officinalis chip	Liu, Min, and Wang 2016a

Table 6. Percentage of wastes generated from different vegetable processing (Gupta and Joshi, 2000)

Source	Nature of waste	Production in tons	Approx. waste in %	Potential quantities of waste (Tons)
Mango	Peel, stones	6987.7	45	3144.4
Banana	Peel	2378.0	35	832.3
Citrus	Peel, rag and seed	1211.9	50	606.0
Pineapple	Skin, core	75.7	33	24.7
Grapes	Stem, skin and seeds	565	20	---
Guavas	Peel and core and seeds	565	10	---
Peas	Shell	107.7	40	68.3
Tomato	Skin, core and seeds	464.5	20	90.3
Potato	Peel	2769.0	15	415.3
Onion	Outer leaves	1102.0	---	---
Apple	Peel, pomace and seeds	1376.0	---	412.0

Source: Gupta and Joshi, 2000

extraction is considered to be the best. Utilization of such techniques will not only save agro wastes being exploited, but also will be economically friendly approach towards healthy steps instead of using chemically synthesized agents in formulation of functional foods. This will lessen the burden of dumping wastes that is also an eco friendly move towards a healthy public domain. In countries, such as India, agro wastes are much higher and waste management is really a tough job ahead. Potential use of these wastes can nullify the hazardous effect towards environment. Agro wastes also showed promising effect in dairy fermentation. Further more knowledge in the fermentation processes in these wastes and the mechanisms would be really helpful to understand their multifunctional role. PVA nano- composites containing cellulose nano- crystals and silver nano- particles are need to be explored more to understand its possible contribution in food processing or any other dimension. Finding out the natural source to be used as adsorbent is also challenging and it is released the button of activation is being pressed leading to drop of temperature to 30°C within a few minutes. Role in water treatment, strengthen properties, cosmetics are also needed to be explored more. Successful water treatment utilizing natural resources will cut the problem regarding unavailability of drinking water along with

Table 7. various bioactive compounds from agricultural food waste residues

Sources	Residues	Bioactive compounds	Reference
Apple	Peel and pomace	Epicatechin, catechins, anthocyanins, quercitin glycosides, chlorogenic acid, hydroxycinnamates, phloretin glycosides and procyanidins	Wolfe and Liu (2003), Foo and Lu (1999), Lu and Foo (1997)
Avocado	Peel and seeds	Epicatechin, catechin, gallic acid, chlorogenic acid, cyanidin 3-glucoside and homogentisic acid	Deng et al. (2012)
Banana	Peel	Gallocatechin, anthocyanins, delphindin, cyaniding, catecholamine	Someya et al. (2002), Kanazawa and Sakakibara(2000), González Montelongo et al. (2010)
Citrus fruits	Peel	Hesperidin, naringin, eriocitrin, narirutin	Coll et al. (1998)
Grapes	Seed and skin	Coumaric acid, cafeic acid, ferulic acid, chlorogenic acid, cinnamic acid, neochlorogenic acid, *p*-hydroxybenzoic acid, protocatechuic acid, vanillic acid, gallic acid, proanthocyanidins, quercetin 3-*o*-gluuronide, quercetin, resvaratrol	Shrikhande (2000), Negro et al. (2003), Maier et al. (2009)
Guava	Skin and seeds	Catechin, cyanidin 3-glucoside, galangin, gallic acid, homogentisic acid, kaempferol	Deng et al. (2012)
Litchi	Pericarp, seeds	Cyanidin-3-glucoside, cyanidin-3-rutonoside, malvidin-3-glucoside, gallic acid, epicatechin-3-gallate	Lee and Wicker (1991), Duan et al. (2007)
Mango	Kernel	Gallic acid, ellagic acid, gallates, gallotannins, condensed tannins	Arogba (2000), Puravankara et al. (2000)
Palm	By-products of palm oil milling	Tocopherols, tocotrienols, sterols, and squalene, phenolic antioxidants	Tan et al. (2007), Choo et al. (1996)
Pomegranate	Peel and pericarp	Gallic acid, cyanidin-3,5-diglucoside, cyanidin-3-diglucoside, delphinidin-3,5-diglucoside	Noda et al. (2002), Gil et al. (2000)
Carrot	Peel	Phenols, beta-carotene	Chantaro et al. (2008)
Cucumber	Peel	Chlorophyll, pheophytin, phellandrene, caryophyllene	Zeyada et al. (2008)
Tomato	Skin and pomace	Carotenoids	Strati and Oreopoulou (2011)
Barley	Bran	β-Glucan	Sainvitu et al. (2012)
Rice	bran	γ-Oryzanol, bran oil	Perretti et al. (2003), Oliveira et al. (2012)
Wheat	Bran and germs	Phenolic acids, antioxidants	Wang et al. (2008)

waste management. Government or any other bodies should arrange campaign or education program to make normal people aware of these untapped properties of agricultural wastes and their beneficial contributions.

CONCLUSION

Maintaining a proper balance between food and hunger issues along with the increasing global population is a tough work to go on smoothly. Rate of produced crops are not sufficient to slow down the hunger problem. Beside this major thing, global population is also an alarming phase. Toxic side effects of synthetic drug for the treatment is reverting back causing other fatal diseases. Use of made up agents for water treatment creates economical issues. All these problems might be decelerated by utilizing the agricultural wastes which are also indirectly causing pollution due to dumping the excessive amount of unused portion of them in open air. These agro wastes are high rich source of nutrients and the untapped health beneficial properties should be reserved for better future purposes. These are generally used as livestock feed and fertilizer. But most of these are remained unused. Formulation of value added food products utilizing these wastes is not only a healthy approach but also helps in waste management. Moreover based upon ongoing studies, the extraction of the agro wastes also possesses various phyto chemicals, bioactive molecules (phenol, antioxidant, anti inflammatory agent) that can be used for therapeutic purposes. This value added food products instead of using synthetic chemicals will help to avoid side effects of the later and also will be promising to combat malnutrition and nutrient deficiencies along with maintenance of food security. Ashes and other discharges, after dumping these agricultural wastes, contains highly toxic contents, such as, silica that can cause fatal diseases like, skin irritation, breathing problem, digestive abnormality etc. Utilization of such wastes can be helpful towards avoiding environmental hazards. Not only food product development and therapeutic value are its sole utility, but also can be an important resource for dairy industry as they proved to be promising in case of fermentation. This natural source can also be used for water treatment leading to economically friendly approach towards availability of drinking water. Its adsorbent property can also be utilized to remove air pollutants. Their enormous strengthen property is also helpful as a replacement of other concrete elements. Through different innovative techniques various beneficial enzyme extractions is also helpful. To explore different dimensions of its properties different technologies are also being improvised which are cutting down the extra cost from the early budget. Based upon these novel technologies, low cost industries can be launched that indirectly may provide financial security to below poverty population of society. This concept of utilizing agricultural wastes is so promising and convincing that developed as well as developing countries is really interested to invest in this eco friendly and beneficial approach towards public domain. Researchers are also showing keen interest and they believe that these natural resources are yet to explore many other crucial properties. Agricultural wastes are not only a source to secure their untapped nutrients but also a crucial agent for other purposes that are beneficial in other way, such as, thermo insulator, adsorbent, building material, nano composite, antibiotic, anti inflammatory agent, antioxidant, food additives and so on. Proper utilization of these natural resources are very much needed to avoid hunger, malnutrition, pollution, chemical side effects to maintain a healthy pollution free life style.

REFERENCES

Abul-Fadl, M. M. (2012). Nutritional and chemical evaluation of white cauliflower by-products flour and the effect of its addition on beef sausage quality. *Journal of Applied Sciences Research*, (2): 693–704.

Adámez, J. D., Samino, E. G., Sánchez, E. V., & González, G. D. (2012). In vitro estimation of the antibacterial activity and antioxidant capacity of aqueous extracts from grape-seeds (*Vitis vinifera* L.). *Food Control*, *24*(1-2), 136–141. doi:10.1016/j.foodcont.2011.09.016

Ajikumar, P. K., Tyo, K., Carlsen, S., Mucha, O., Phon, T. H., & Stephanopoulos, G. (2008). Terpenoids: Opportunities for biosynthesis of natural product drugs using engineered microorganisms. *Molecular Pharmaceutics*, *5*(2), 167–190. doi:10.1021/mp700151b PMID:18355030

Akpan, I., Bankole, M. O., Adesemowo, A. M., & Latunde, D. G. (1999). Production of amylase by *A. niger* in a cheap solid medium using rice bran and agricultural materials. *Tropical Science*, *39*, 77–79.

Aliakbarian, B., Fathi, A., Perego, P., & Dehghani, F. (2012). Extraction of antioxidants from winery wastes using subcritical water. *The Journal of Supercritical Fluids*, *65*(65), 18–24. doi:10.1016/j.supflu.2012.02.022

Arogba, S. S. (2000). Mango (*Mangifera indica*) kernel: Chromatographic analysis of the tannin, and stability study of the associated polyphenol oxidase activity. *Journal of Food Composition and Analysis*, *13*(2), 149–156. doi:10.1006/jfca.1999.0838

Ayala-Zavala, J., Rosas-Domínguez, C., Vega-Vega, V., & González-Aguilar, G. (2010). Antioxidant enrichment and antimicrobial protection of fresh-cut fruits using their own by products: Looking for integral exploitation. *Journal of Food Science*, *75*(8), R175–R181. doi:10.1111/j.1750-3841.2010.01792.x PMID:21535513

Ayala-Zavala, J. F., González-Aguilar, G. A., & Del-Toro-Sánchez, L. (2009). Enhancing safety and aroma appealing of fresh-cut fruits and vegetables using the antimicrobial and aromatic power of essential oils. *Journal of Food Science*, *74*(7), R84–R91. doi:10.1111/j.1750-3841.2009.01294.x PMID:19895494

Baiano, A. (2014). Recovery of biomolecules from food wastes-a review. *Molecules (Basel, Switzerland)*, *19*(9), 14821–14842. doi:10.3390/molecules190914821 PMID:25232705

Barbosa, H. M. A., Melo, M. M. R. D., Coimbra, M. A., Passos, C. P., & Silva, C. M. (2014). Optimization of the supercritical fluid co-extraction of oil and diterpenes from spent coffee grounds using experimental design and response surface methodology. *The Journal of Supercritical Fluids*, *85*(1), 165–172. doi:10.1016/j.supflu.2013.11.011

Barrales, F. M., Rezende, C. A., & Martınez, J. (2015). Supercritical CO2 extraction of passion fruit (Passiflora edulis sp.) seed oil assisted by ultrasound. *The Journal of Supercritical Fluids*, *104*, 183–192. doi:10.1016/j.supflu.2015.06.006

Belewu, M. A., & Babalola, F. T. (2009). Nutrient enrichment of some waste agricultural residues after solid state fermentation using *Rhizopus oligosporus*. *J Appl Biosci*, *13*, 695–699.

Benazzi, T., Calgaroto, S., Rosa, C. D., Oliveira, J. V., & Mazutti, M. A. (2013). Hydrolysis of sugarcane bagasse using supercritical carbon dioxide to obtain fermentable sugars. *Journal of Chemical Technology and Biotechnology (Oxford, Oxfordshire)*, *88*(9), 1766–1768. doi:10.1002/jctb.4002

Cagdas, E., & Kumcuoglu, S. (2015). Effect of grape seed powder on oxidative stability of precooked chicken nuggets during frozen storage. *Journal of Food Science and Technology*, *52*(5), 2918–2925. doi:10.100713197-014-1333-7 PMID:25892791

Cai, L., Choi, I., Hyun, J. N., Jeong, Y. K., & Baik, B. K. (2014). Influence of bran particle size on bread-baking quality of whole grain wheat flour and starch retrogradation. *Cereal Chemistry*, *91*(1), 65–71. doi:10.1094/CCHEM-02-13-0026-R

Casas Cardoso, L., Mantell Serrano, C., Torrez Quintero, E., Pereyra Lopez, C., Medrano Antezana, R., & Martinez de la Ossa, E. J. (2013). High pressure extraction of antioxidants from solanum stenotomun peel. *Molecules (Basel, Switzerland), 18*(3), 3137–3151. doi:10.3390/molecules18033137 PMID:23528996

Chantaro, P., Devahastin, S., & Chiewchan, N. (2008). Production of antioxidant high dietary fber powder from carrot peels. *Lebensmittel-Wissenschaft + Technologie, 41*(10), 1987–1994. doi:10.1016/j.lwt.2007.11.013

Choo, Y.-M., Yap, S.-C., Ooi, C.-K., Ma, A.-N., Goh, S.-H., & Ong, A. S.-H. (1996). Recovered oil from palm-pressed fber: A good source of natural carotenoids, vitamin E, and sterols. *Journal of the American Oil Chemists' Society, 73*(5), 599–602. doi:10.1007/BF02518114

Chouchouli, V., Kalogeropoulos, N., Konteles, S. J., Karvela, E., Makris, D. P., & Karathanos, V. T. (2013). Fortification of yoghurts with grape (Vitis vinifera) seed extracts. *Lebensmittel-Wissenschaft + Technologie, 53*(2), 522–529. doi:10.1016/j.lwt.2013.03.008

Coll, M., Coll, L., Laencina, J., & Tomas-Barberan, F. (1998). Recovery of favanones from wastes of industrially processed lemons. *European Food Research and Technology, 206*, 404–407.

Crizel, Jablonski, Rios, Rech, & Flores. (2013). Dietary fiber from orange byproducts as a potential fat replacer. *Lwt-Food Science and Technology, 53*(1), 9–14. doi:.lwt.2013.02.002 doi:10.1016/j

Daifullah, A., Girgis, B., & Gad, H. (2004). A study of the factors affecting the removal of humic acid by activated carbon prepared from biomass material. *Colloids and Surfaces. A, Physicochemical and Engineering Aspects, 235*(1-3), 1–10. doi:10.1016/j.colsurfa.2003.12.020

Dasgupta, J., Kumar, A., Mandal, D. D., Mandal, T., & Datta, S. (2015). Removal of phenol from aqueous solutions using adsorbents derived from low-cost agro-residues. *Desalination and Water Treatment*. doi:10.1080/19443994.2015.1061455

Deng, G.-F., Shen, C., Xu, X.-R., Kuang, R.-D., Guo, Y.-J., Zeng, L.-S., ... Xia, E.-Q. (2012). Potential of fruit wastes as natural resources of bioactive compounds. *International Journal of Molecular Sciences, 13*(7), 8308–8323. doi:10.3390/ijms13078308 PMID:22942704 do Espirito Santo, A. P., Cartolano, N. S., Silva, T. F., Soares, F. A. S. M., Gioielli, L. A., Perego, P., ... Oliveira, M. N. (2012a). Fibers from fruit by-products enhance probiotic viability and fatty acid profile and increase CLA content in yoghurts. *International Journal of Food Microbiology, 154*(3), 135–144. doi:10.1016/j.ijfoodmicro.2011.12.025 PMID:22264421

do Espirito Santo, A. P., Perego, R., Converti, A., & Oliveira, M. N. (2012b). Influence of milk type and addition of passion fruit peel powder on fermentation kinetics, texture profile and bacterial viability in probiotic yoghurts. *Lebensmittel-Wissenschaft + Technologie, 47*(2), 393–399. doi:10.1016/j.lwt.2012.01.038

Doke, K. M., & Khan, E. M. (2013). Adsorption thermodynamics to clean up wastewater; critical review. *Reviews in Environmental Science and Biotechnology, 12*(1), 25–44. doi:10.100711157-012-9273-z

Duan, X., Jiang, Y., Su, X., Zhang, Z., & Shi, J. (2007). Antioxidant properties of anthocyanins extracted from litchi (*Litchi chinenesis* Sonn.) fruit pericarp tissues in relation to their role in the pericarp browning. *Food Chemistry, 101*(4), 1365–1371. doi:10.1016/j.foodchem.2005.06.057

Ekinci & Gurcu. (2014). Extraction of oil and € b-sitosterol from peach (Prunus persica) seeds using supercritical carbon dioxide. *Journal of Supercritical Fluids, 92*, 319–323. doi:. supflu.2014.06.004 doi:10.1016/j

El-Tayeb, T. S., Abdelhafez, A. A., Ali, S. H., & Ramadan, E. M. (2012). Efect of acid hydrolysis and fungal biotreatment on agro-industrial wastes for obtainment of free sugars for bioethanol production. *Brazilian Journal of Microbiology, 43*(4), 1523–1535. doi:10.1590/S1517-83822012000400037 PMID:24031984

Farıas-Campomanes, Rostagno, & Meireles. (2013). Production of polyphenol extracts from grape bagasse using supercritical fluids: Yield, extract composition and economic evaluation. *Journal of Supercritical Fluids, 77*(5), 70–78. doi:. supflu.2013.02.006 doi:10.1016/j

Foo, K. Y., & Hameed, B. H. (2009a). A short review of activated carbon assisted electrosorption process: An overview, current stage and future prospects. *Journal of Hazardous Materials, 170*(2-3), 552–559. doi:10.1016/j.jhazmat.2009.05.057 PMID:19501461

Foo, K. Y., & Hameed, B. H. (2009c). Utilization of biodiesel waste as a renewable resource for activated mcarbon: Application to environmental problems. *Renewable & Sustainable Energy Reviews, 13*(9), 2495–2504. doi:10.1016/j.rser.2009.06.009

Foo, L. Y., & Lu, Y. (1999). Isolation and identifcation of procyanidins in apple pomace. *Food Chemistry*, *64*(4), 511–518. doi:10.1016/S0308-8146(98)00150-2

Garcıa-Lomillo, Gonzalez-Sanjose, Pino-Garcıa, & Ortega-Heras. (2017). Antioxidant effect of seasonings derived from wine pomace on lipid oxidation in refrigerated and frozen beef patties. *LWT – Food Science and Technology, 77*, 85–91. doi:.2016.11.038 doi:10.1016/j.lwt

Garcia-Lomillo, J., Viegas, O., Gonzalez-SanJose, M. L., & Ferreira, I. M. P. L. V. O. (2016b). Influence of red wine pomace seasoning and highoxygen atmosphere storage on carcinogens formation in barbecued beef patties. *Meat Science*, *125*, 10–15. doi:10.1016/j.meatsci.2016.11.009 PMID:27871036

Gayen, S., & Ghosh, U. (2011). Pectin methyl esterase production from mixed agrowastes by *Penicillium notatum* NCIM 923 in solid state fermentation. *Journal of Bioremediation & Biodegradation*, *2*(02), 119. doi:10.4172/2155-6199.1000119

Ghafoor, K., Al-Juhaimi, F. Y., & Choi, Y. H. (2012). Supercritical fluid extraction of phenolic compounds and antioxidants from grape (Vitis labrusca B.) seeds. *Plant Foods for Human Nutrition (Dordrecht, Netherlands)*, *67*(4), 407–414. doi:10.100711130-012-0313-1 PMID:22992888

Gil, M. I., Tomás-Barberán, F. A., Hess-Pierce, B., Holcroft, D. M., & Kader, A. A. (2000). Antioxidant activity of pomegranate juice and its relationship with phenolic composition and processing. *Journal of Agricultural and Food Chemistry*, *48*(10), 4581–4589. doi:10.1021/jf000404a PMID:11052704

Godfray, H. C. J., Beddington, J. R., Crute, I. R., Haddad, L., Lawrence, D., Muir, J. F., ... Toulmin, C. (2010). Food security: The challenge of feeding 9 billion people. *Science*, *327*(5967), 812–818. doi:10.1126cience.1185383 PMID:20110467

González-Montelongo, R., Lobo, M. G., & González, M. (2010). Antioxidant activity in banana peel extracts: Testing extraction conditions and related bioactive compounds. *Food Chemistry*, *119*(3), 1030–1039. doi:10.1016/j.foodchem.2009.08.012

Gupta, K., & Joshi, V. K. 2000. Fermentative utilization of waste from food processing industry In Postharvest Technology of Fruits and Vegetables: Handling Processing Fermentation and Waste Mangement (vol. 2). Indus Pub Co.

He, L., Zhang, X., Xu, H., Xu, C., Yuan, F., Knez, Z., ... Gao, Y. (2012). Subcritical water extraction of phenolic compounds from pomegranate (Punica granatum L.) seed residues and investigation into their antioxidant activities with HPLC-ABTS(center dotC) assay. *Food and Bioproducts Processing*, *90*(2C2), 215–223. doi:10.1016/j.fbp.2011.03.003

Joana Gil-Chávez, G., Villa, J. A., Fernando Ayala-Zavala, J., Basilio Heredia, J., Sepulveda, D., Yahia, E. M., & González-Aguilar, G. A. (2013). Technologies for extraction and production of bioactive compounds to be used as nutraceuticals and food ingredients: An overview. *Comprehensive Reviews in Food Science and Food Safety*, *12*(1), 5–23. doi:10.1111/1541-4337.12005

Kalapathy, U., Proctor, A., & Shultz, J. (2000). A simple method for production of pure silica from rice hull ash. *Bioresource Technology*, *73*(3), 257–262. doi:10.1016/S0960-8524(99)00127-3

Kalogeropoulos, N., Chiou, A., Pyriochou, V., Peristeraki, A., & Karathanos, V. T. (2012). Bioactive phytochemicals in industrial tomatoes and their processing byproducts. *Lebensmittel-Wissenschaft + Technologie*, *49*(2), 213–216. doi:10.1016/j.lwt.2011.12.036

Kanatt, S. R., Arjun, K., & Sharma, A. (2011). Antioxidant and antimicrobial activity of legume hulls. *Food Research International*, *44*(10), 3182–3187. doi:10.1016/j.foodres.2011.08.022

Kanazawa, K., & Sakakibara, H. (2000). High content of dopamine, a strong antioxidant, in cavendish banana. *Journal of Agricultural and Food Chemistry*, *48*(3), 844–848. doi:10.1021/jf9909860 PMID:10725161

Katalinic, V., Mozina, S. S., Skroza, D., Generalic, I., Abramovic, H., Milos, M., ... Boban, M. (2010). Polyphenolic profle, antioxidant properties and antimicrobial activity of grape skin extracts of 14 *Vitis vinifera* varieties grown in dalmatia (Croatia). *Food Chemistry*, *119*(2), 715–723. doi:10.1016/j.foodchem.2009.07.019

Kaur, G., Sharma, S., Nagi, H. P. S., & Dar, B. N. (2012). Functional properties of pasta enriched with variable cereal brans. *Journal of Food Science and Technology-Mysore*, *49*(4), 467–474. doi:10.100713197-011-0294-3 PMID:23904655

Kitinoja, L., Saran, S., Roy, S. K., & Kader, A. A. (2011). Postharvest technology for developing countries: Challengesand opportunities in research, outreach and advocacy. *Journal of the Science of Food and Agriculture*, *91*(4), 597–603. doi:10.1002/jsfa.4295 PMID:21302312

Ko, Cheigh, Cho, & Chung. (2011). Subcritical water extraction of flavonol quercetin from onion skin. *Journal of Food Engineering*, *102*(4), 327–333. doi:.09.008 doi:10.1016/j.jfoodeng.2010

Krishna, G. R., Rao, D. S., Rao, Z. P., & Prasad, J. R. (2002). Nutrient utilization of concentrate mixtures with varying levels of green gram (Vigna radiata) chuni by native male buffaloes. *Buffalo Journal*, *18*(1), 59–69.

Kumar, A., Sengupta, B., Dasgupta, D., Mandal, T., & Datta, S. (2015). Recovery of value added products from rice husk ash to explore an economic way for recycle and reuse of agricultural waste. *Reviews in Environmental Science and Biotechnology*.

Kumar, P., Prasad, B., & Chand, S. (2009). Treatment of desizing wastewater by catalytic thermal treatment and coagulation. *Journal of Hazardous Materials*, *163*(1), 433–440. doi:10.1016/j.jhazmat.2008.06.114 PMID:18687524

Lee, H., & Wicker, L. (1991). Anthocyanin pigments in the skin of lychee fruit. *Journal of Food Science*, *56*(2), 466–468. doi:10.1111/j.1365-2621.1991.tb05305.x

Lina, F., Ballesteros, J. A., & Teixeira, S. I. (2014). Chemical, functional, and structural properties of spent coffee grounds and cofee silverskin. *Food and Bioprocess Technology*, *7*(12), 3493–3503. doi:10.100711947-014-1349-z

Liu, Z., Min, Z., & Wang, Y. (2016). Drying of restructured chips made from the old stalks of Asparagus officinalis: Impact of different drying methods. *Journal of the Science of Food and Agriculture*, *96*(8), 2815–2824. doi:10.1002/jsfa.7449 PMID:26337496

Lu, Y., & Foo, L. Y. (1997). Identifcation and quantifcation of major polyphenols in apple pomace. *Food Chemistry*, *59*(2), 187–194. doi:10.1016/S0308-8146(96)00287-7

Machmudah, Zakaria, Winardi, Sasaki, Goto, Kusumoto, & Hayakawa. (2012). Lycopene extraction from tomato peel byproduct containing tomato seed using supercritical carbon dioxide. *Journal of Food Engineering*, *108*(2), 290–296. doi:.2011.08.012 doi:10.1016/j.jfoodeng

Maier, T., Schieber, A., Kammerer, D. R., & Carle, R. (2009). Residues of grape (*Vitis vinifera* L.) seed oil production as a valuable source of phenolic antioxidants. *Food Chemistry*, *112*(3), 551–559. doi:10.1016/j.foodchem.2008.06.005

Manna, L., Bugnone, C. A., & Banchero, M. (2015). Valorization of hazelnut, coffee and grape wastes through supercritical fluid extraction of triglycerides and polyphenols. *The Journal of Supercritical Fluids*, *104*, 204–211. doi:10.1016/j.supflu.2015.06.012

Marques, J. L., Porta, G. D., Reverchon, E., Renuncio, J. A. R., & Mainar, A. M. (2013). Supercritical antisolvent extraction of antioxidants from grape seeds after vinification. *The Journal of Supercritical Fluids*, *82*(10), 238–243. doi:10.1016/j.supflu.2013.07.005

Martin, J. G. P., Porto, E., Correa, C. B., Alencar, S. M., Gloria, E. M., Cabral, I. S. R., & Aquino, L. M. (2012). Antimicrobial potential and chemical composition of agroindustrial wastes. *Journal of Natural Products*, *5*, 27–36.

Mehta, K., & Duhan, J. S. (2014). Production of invertase from *Aspergillus niger* using fruit peel waste as a substrate. *International Journal of Pharma and Bio Sciences*, *5*(2), B353–B360.

Melo, M. M. R. D., Barbosa, H. M. A., Passos, C. P., & Silva, C. M. (2014). Supercritical fluid extraction of spent coffee grounds: Measurement of extraction curves, oil characterization and economic analysis. *The Journal of Supercritical Fluids*, *86*(1), 150–159. doi:10.1016/j.supflu.2013.12.016

Motte, J. C., Trably, E., Escudié, R., Hamelin, J., Steyer, J. P., Bernet, N., ... Dumas, C. (2013). Total solids content: A key parameter of metabolic pathways in dry anaerobic digestion. *Biotechnology for Biofuels*, *6*(1), 164. doi:10.1186/1754-6834-6-164 PMID:24261971

Negro, C., Tommasi, L., & Miceli, A. (2003). Phenolic compounds and antioxidant activity from red grape marc extracts. *Bioresource Technology*, *87*(1), 41–44. doi:10.1016/S0960-8524(02)00202-X PMID:12733573

Nguyen, T. A. D., Kim, K. R., Han, S. J., Cho, H. Y., Kim, J. W., Park, S. M., & Sim, S. J. (2010). Pretreatment of rice straw with ammonia and ionic liquid for lignocelluloses conversion to fermentable sugars. *Bioresource Technology*, *101*(19), 7432–7438. doi:10.1016/j.biortech.2010.04.053 PMID:20466540

Nigam, P. S., Gupta, N., & Anthwal, A. (2009). Pre-treatment of agro-industrial residues. In P. S. Nigam & A. Pandey (Eds.), *Biotechnology for agro-industrial residues utilization* (pp. 13–33). Heidelberg, Germany: Springer. doi:10.1007/978-1-4020-9942-7_2

Nobre, B. P., Gouveia, L., Matos, P. G. S., Cristino, A. F., Palavra, A. F., & Mendes, R. L. (2012b). Supercritical extraction of lycopene from tomato industrial wastes with ethane. *Molecules (Basel, Switzerland)*, *17*(7), 8397–8407. doi:10.3390/molecules17078397 PMID:22785267

Noda, Y., Kaneyuki, T., Mori, A., & Packer, L. (2002). Antioxidant activities of pomegranate fruit extract and its anthocyanidins: Delphinidin, cyanidin, and pelargonidin. *Journal of Agricultural and Food Chemistry*, *50*(1), 166–171. doi:10.1021/jf0108765 PMID:11754562

Oliveira, R., Oliveira, V., Aracava, K. K., & da Costa Rodrigues, C. E. (2012). Efects of the extraction conditions on the yield and composition of rice bran oil extracted with ethanol—A response surface approach. *Food and Bioproducts Processing*, *90*(1), 22–31. doi:10.1016/j.fbp.2011.01.004

Paepatung, N., Nopharatana, A., & Songkasiri, W. (2009). Bio-methane potential of biological solid materials and agricultural wastes. *Asian J Energy Env*, *10*, 19–27.

Parfitt, J., Barthel, M., & Macnaughton, S. (2010). Food waste within food supply chains: Quantification and potentialfor change to 2050. Philos. Trans. R. Soc. B Biol. *Sci.*, *365*, 3065–3081. PMID:20713403

Patel, M. (2012). Potential of fruit and vegetable wastes as novel biosorbents: Summarizing the recent studies. *Reviews in Environmental Science and Biotechnology*, *11*(4), 365–380. doi:10.100711157-012-9297-4

Perretti, G., Miniati, E., Montanari, L., & Fantozzi, P. (2003). Improving the value of rice by-products by SFE. *The Journal of Supercritical Fluids*, *26*(1), 63–71. doi:10.1016/S0896-8446(02)00247-4

Prado, J. M., Forster-Carneiro, T., Rostagno, M. A., Follegatti-Romero, L. A., Maugeri Filho, F., & Meireles, M. A. A. (2014). Obtaining sugars from coconut husk, defatted grape seed, and pressed palm fiber by hydrolysis with subcritical water. *The Journal of Supercritical Fluids*, *89*, 89–98. doi:10.1016/j.supflu.2014.02.017

Puravankara, D., Boghra, V., & Sharma, R. S. (2000). Efect of antioxidant principles isolated from mango (*Mangifera indica* L) seed kernels on oxidative stability of bufalo ghee (butter-fat). *Journal of the Science of Food and Agriculture*, *80*(4), 522–526. doi:10.1002/(SICI)1097-0010(200003)80:4<522::AID-JSFA560>3.0.CO;2-R

Rababah, T. M., Yucel, S., Ereifej, K. I., Alhamad, M. N., Al-Mahasneh, M. A., Yang, W., ... Ismaeal, K. (2011). Effect of grape seed extracts on the physicochemical and sensory properties of corn chips during storage. *Journal of the American Oil Chemists' Society*, *88*(5), 631–637. doi:10.100711746-010-1709-4

Ramachandran, S., Patel, A. K., Nampoothiri, K. M., Francis, F., Nagy, V., Szakacs, G., & Pandey, A. (2004). Coconut oil cake—A potential raw material for the production of a-amylase. *Bioresource Technology*, *93*(2), 169–174. doi:10.1016/j.biortech.2003.10.021 PMID:15051078

Reategui, Machado, Barbero, Rezende, & Martınez. (2014). Extraction of antioxidant compounds from blackberry (Rubus sp.) bagasse using supercritical CO2 assisted by ultrasound. *Journal of Supercritical Fluids, 94*, 223–233. doi:.supflu.2014.07.019 doi:10.1016/j

Rekha, K. S. S., Lakshmi, C., Devi, S. V., & Kumar, M. S. (2012). Production and optimization of lipase from *Candida rugosa* using groundnut oilcake under solid state fermentation. *Intern J Res Eng Technol, 1*(04), 571–577. doi:10.15623/ijret.2012.0104008

Rivas, B., Torrado, A., Torre, P., Converti, A., & Domínguez, J. M. (2008). Submerged citric acid fermentation on orange peel autohydrolysate. *Journal of Agricultural and Food Chemistry, 56*(7), 2380–2387. doi:10.1021/jf073388r PMID:18321055

Rudra, S. G., Nishad, J., Jakhar, N., & Kaur, C. (2015). Food industry waste: Mine of nutraceuticals. *Int J Sci Environ Technol, 4*, 205–229.

Sainvitu, P., Nott, K., Richard, G., Blecker, C., Jérôme, C., Wathelet, J.-P., ... Deleu, M. (2012). Structure, properties and obtention routes of faxseed lignin secoisolariciresinol: A review. *Biotechnologie, Agronomie, Société et Environnement, 16*, 115.

Sant'Anna, V., Dalla Porta Christiano, F., Ferreira Marczak, L. D., Tessaro, I. C., & Silveira Thys, R. C. (2014). The effect of the incorporation of grape marc powder in fettuccini pasta properties. *Lebensmittel-Wissenschaft + Technologie, 58*(2), 497–501. doi:10.1016/j.lwt.2014.04.008

Santos, D. T., Albarelli, J. Q., Beppu, M. M., & Meireles, M. A. A. (2013). Stabilization of anthocyanin extract from jabuticaba skins by encapsulation using supercritical CO 2 as solvent. *Food Research International, 50*(2), 617–624. doi:10.1016/j.foodres.2011.04.019

Santos, D. T., Veggi, P. C., & Meireles, M. A. A. (2012). Optimization and economic evaluation of pressurized liquid extraction of phenolic compounds from jabuticaba skins. *Journal of Food Engineering, 108*(3), 444–452. doi:10.1016/j.jfoodeng.2011.08.022

Sharanappa, A., Wani, K. S., & Pallavi, P. (2011). Bioprocessing of food industrial waste for α-amylase production by solid state fermentation. *Intern J Adv Biotechnol Res, 2*, 473–480.

Shi, W., Wang, S., & Yang, Q. (2010). Climate change and global warming. *Reviews in Environmental Science and Biotechnology, 9*(2), 99–102. doi:10.100711157-010-9206-7

Shrikhande, A. J. (2000). Wine by-products with health benefts. *Food Research International, 33*(6), 469–474. doi:10.1016/S0963-9969(00)00071-5

Sindiri, M. K., Machavarapu, M., & Vangalapati, M. (2013). Alfa-amylase production and purifcation using fermented orange peel in solid state fermentation by *Aspergillus niger. Ind J Appl Res*, *3*(8), 49–51. doi:10.15373/2249555X/AUG2013/16

Singh, P. P., & Mda, S. A. (2011). Subcritical water extraction of phenolic compounds from potato peel. *Food Research International*, *44*(8), 2452–2458. doi:10.1016/j.foodres.2011.02.006

Sodhi, H. K., Sharma, K., Gupta, J. K., & Soni, S. K. (2005). Production of a thermostable a-amylase from *Bacillus* sp. PS-7 by solid-state fermentation and its synergistic use in the hydrolysis of malt starch for alcohol production. *Process Biochemistry (Barking, London, England)*, *40*(2), 525–534. doi:10.1016/j.procbio.2003.10.008

Someya, S., Yoshiki, Y., & Okubo, K. (2002). Antioxidant compounds from bananas (*Musa cavendish*). *Food Chemistry*, *79*(3), 351–354. doi:10.1016/S0308-8146(02)00186-3

Song, Y., Ding, S., Chen, S., Xu, H., Mei, Y., & Ren, J. (2015). Removal of malachite green in aqueous solution by adsorption on sawdust. *Korean Journal of Chemical Engineering*, *32*(12), 2443–2448. doi:10.100711814-015-0103-1

Strati, I. F., & Oreopoulou, V. (2011). Efect of extraction parameters on the carotenoid recovery from tomato waste. *International Journal of Food Science & Technology*, *46*(1), 23–29. doi:10.1111/j.1365-2621.2010.02496.x

Tan, Y. A., Sambanthamurthi, R., Sundram, K., & Wahid, M. B. (2007). Valorisation of palm by-products as functional components. *European Journal of Lipid Science and Technology*, *109*(4), 380–393. doi:10.1002/ejlt.200600251

Tseng, A., & Zhao, Y. (2013). Wine grape pomace as antioxidant dietary fibre for enhancing nutritional value and improving storability of yogurt and salad dressing. *Food Chemistry*, *138*(1), 356–365. doi:10.1016/j.foodchem.2012.09.148 PMID:23265499

Turning agriculture waste into new industries. (2018, August 24). Retrieved 1 June 2019 from https://phys.org/news/2018-08-agriculture-industries.html

Walker, R., Tseng, A., Cavender, G., Ross, A., & Zhao, Y. (2014). Physicochemical, nutritional, and sensory qualities of wine grape pomace fortified baked goods. *Journal of Food Science*, *79*(9), S1811–S1822. doi:10.1111/1750-3841.12554 PMID:25102950

Wang, J., Sun, B., Cao, Y., Tian, Y., & Li, X. (2008). Optimisation of ultrasound-assisted extraction of phenolic compounds from wheat bran. *Food Chemistry*, *106*(2), 804–810. doi:10.1016/j.foodchem.2007.06.062

Weshahy, A. A., & Rao, V. A. (2012). Potato peel as a source of important phytochemical antioxidant nutraceuticals and their role in human health—a review. Phytochemicals as nutraceuticals—Global approaches to their role in nutrition and health, 207–224.

Wolfe, K. L., & Liu, R. H. (2003). Apple peels as a value-added food ingredient. *Journal of Agricultural and Food Chemistry*, *51*(6), 1676–1683. doi:10.1021/jf025916z PMID:12617604

Yi, T., Huang, X., Pan, S., & Wang, L. (2014). Physicochemical and functional properties of micronized jincheng orange by-products (Citrus sinensis Osbeck) dietary fiber and its application as a fat replacer in yogurt. *International Journal of Food Sciences and Nutrition*, *65*(5), 565–572. doi:10.3109/09637486.2014.898252 PMID:24625022

Yilmaz, E. E., Ozvural, E. B., & Vural, H. (2011). Extraction and identification of proanthocyanidins from grape seed (Vitis Vinifera) using supercritical carbon dioxide. *The Journal of Supercritical Fluids*, *55*(3), 924–928. doi:10.1016/j.supflu.2010.10.046

Zafar, S. (2014). *Waste Management, waste- to- energy*. Academic Press.

Zeyada, N. N., Zeitoum, M., & Barbary, O. (2008). Utilization of some vegetables and fruit waste as natural antioxidants. *Alexandria Journal of Food Science and Technology*, *5*, 1–11.

Zhang, H., Ding, X., Chen, X., Ma, Y., Wang, Z., & Zhao, X. (2015). A new method of utilizing rice husk: Consecutively preparing D-xylose, organosolv lignin, ethanol and amorphous superfine silica. *Journal of Hazardous Materials*, *291*, 65–73. doi:10.1016/j.jhazmat.2015.03.003 PMID:25768989

Zhao, Lu, Wu, Liu, & Wang. (2012). Optimization of supercritical phase and combined supercritical/subcritical conversion of lignocellulose for hexose production by using a flow reaction system. *Bioresource Technology, 126*(12), 391–396. doi:. biortech.2012.03.001 doi:10.1016/j

Zhu, G. Y., Xiao, Z. B., Xian, Z., Yi, F. P., & Wan, X. L. (2013). Reducing sugars production from sugarcane bagasse wastes by hydrolysis in subcritical water. *Clean Technologies and Environmental Policy*, *15*(1), 55–61. doi:10.100710098-012-0476-3

KEY TERMS AND DEFINITIONS

Agricultural Waste: Waste that are produced from various agricultural operations. It mainly includes manure and other wastes from farms, poultry houses and slaughterhouses, harvest waste, fertilizer run-off from fields, pesticides which enter into water, air or soils; and salt and silt drained from fields.

Antioxidants: Compounds that inhibit oxidation which is a chemical reaction that can produce free radicals (ROS), leading to chain reactions that may damage the cells of organisms. Antioxidants viz., thiols or ascorbic acid cease these chain reactions.

Biodegradation: The conversion of a substance into other compounds through biochemical reactions and/or the actions of microorganisms such as bacteria is known as biodegradation. Alternatively, it can be defined as a process by which microbial organisms transform or alter (through metabolic or enzymatic action) the structure of chemicals.

Fermentation: A metabolic process that produces chemical changes in organic substrates through the action of enzymes. In bio- chemistry, it is narrowly defined as the extraction of energy from carbohydrates in the absence of oxygen.

Food Security: Food security, as defined by the United Nations' Committee on World Food Security, means that all people, at all times, have physical, social, and economic access to sufficient, safe, and nutritious food that meets their food preferences and dietary needs for an active and healthy life.

Functional Food: A food with an additional function due to accumulation of new ingredients or more of existing ingredients.

Industrial Waste: Waste that are produced due to industrial activity, including any material that is rendered useless during a manufacturing process such as that of factories, industries, mills, and mining operations.

Nutraceuticals: Any product derived from food sources with extra health beneficial properties in addition to the basic nutritional value found in foods.

Nutrition Security: The consumption of an ample range of foods which provides the essential needed nutrients.

Phyto Chemicals: Chemical compounds that mainly come from plant sources, generally with the purpose of self-survival from competitors, predators, or pathogens are phytochemicals. These chemicals are found be health beneficial. Some phyto chemicals have been used as poisons and others as traditional medicine.

Value Addition: Extra value added over the original value of something. It can apply to products, services, companies, management, as well as other areas of business. In other words, value-added is the enhancement made by a company/individual to a product or service before channelizing it before consumer.

Chapter 7

Dairy Waste Treatment and Utilization:
A Treatise

Anum Ishaq
Department of Allied Health Sciences, Superior College, University Campus, Lahore, Pakistan

Ubaid ur Rahman
Department of Allied Health Sciences, Superior College, University Campus, Lahore, Pakistan

Muhammad Haseeb Ahmad
National Institute of Food Science and Technology, University of Agriculture, Faisalabad, Pakistan

Amna Sahar
https://orcid.org/0000-0002-9299-4000
National Institute of Food Science and Technology, University of Agriculture, Faisalabad, Pakistan

Qamar Abbas Syed
National Institute of Food Science and Technology, University of Agriculture, Faisalabad, Pakistan

Rashida Perveen
Department of Allied Health Sciences, Superior College, University Campus, Lahore, Pakistan

DOI: 10.4018/978-1-7998-1966-0.ch007

ABSTRACT

The dairy industry produces waste materials and by-products including soluble and suspended organic substances, wastewater, suspended solids, and whey. These substances contain significant amounts of organic and inorganic toxins that can cause serious health threats if not treated properly. Additionally, toxins produced by dairy waste would negatively affect the environment as well as quality and availability of aquatic resources. However, these waste materials and by-products can be utilized in different ways after proper treatment and processing. This chapter will provide an overview of waste products of dairy industries and highlight different treatments and utilization of these by-products in other industries.

INTRODUCTION

Dairy industry has been growing tremendously both in size and number all over the world. These industries discharge wastewater characterized by nutritional contents, and inorganic and organic substances (Ahmad *et al.* 2019; Martin-Rilo *et al.* 2015). These wastewaters extensively pollute water bodies if discharged without any treatment (Kubota and da Rosa 2013). The dairy industrial wastes are usually generated as a result of washing and cleaning (Willers *et al.* 2014). According to a rough estimate, around 2% milk is subjected to wastage during milk processing. Dairy waste generally contains lactose, fats, and other nutrients along with several detergents and sanitizing agents. Dairy effluents are rapidly decomposed resulting in depletion of oxygen level and generation of anaerobic conditions in the receiving streams. This induces the production of strong off-odors due to nuisance conditions. It is also revealed that greater concentration of milk waste materials is harmful to marine life. For intense, sometimes casein precipitates in waste water and further transformed into blackish odorous material which is harmful for aquatic life. Dairy waste also possesses plentiful amount of dissolvable organics, other trace organic material and suspended solutions.

Wastewater generated in milk plants contains high concentrations of inorganic (nitrites, nitrates, ammonium salts etc.) and organic (organic carbon, hydrogen gas, waste water) pollutants and its treatment is a complicated task and requires considerable financial funds (Chong *et al.* 2009; Demirel *et al.* 2005). Milk plant wastewater treatment is conducted with the use of biological, chemical and mechanical methods. Biological methods involve aerobic and anaerobic digestion (World Bank 1998). The disadvantages of aerobic biological wastewater treatment technologies of milk plants are, among others, the high costs of electricity in the aeration tanks of air supply; and rate filters. In addition to that, elevated concentration of organic waste material can also lead towards the activation of sludge and induce serious problems in working of secondary sedimentation tanks.

Cheese whey is one of the major pollutants of dairy industry which is produced after the separation of casein and fat from the milk. Indeed, whey contains significant amount of carbon and nitrogenous compounds that can be used as substrate in different fermentation processes. Bioactive whey proteins and peptides can also be used in different food products, animal feed, pharmaceuticals and as functional compounds (Marshall 2004).

Several investigations have provided information regarding the treatment and utilization of dairy effluents. Waste generated from dairy processing industry can be treated by using various physico-chemical and biological methods such as coagulation/flocculation, wetland treatment, use of trickling filters, anaerobic filters, activated sludge treatment, aerated lagoons, sequential batch reactor, up-flow anaerobic sludge blanket, aerobic and anaerobic fermentation, anaerobic digestion etc. By using these treatment methods, several by-products can be developed such as bioplastics, biofuels, single cell proteins, organic acids, polysaccharides, bioactive peptides, biosurfactants etc. Many of these by-products are quite useful in developing various health-promoting formulations in food and pharmaceutical industries. Accordingly, the aim of this chapter is to provide information regarding different types of effluents generated by dairy industry and ways to treat the generated effluents in an effective manner for their proper utilization.

WASTES GENERATED BY DAIRY INDUSTRY

The composition, volume and concentration of dairy effluents depends on the production process, type of product, methods used, plant design, extent of water management and the amount of water being conserved. The major wastes of dairy industry include waste water and whey. Characteristics of these effluents and treatment of wastes is discussed in the following lines.

Dairy Waste Water

Dairy wastewater is further divided into three categories: processing water, cleaning waste water and sanitary waste water. the former contains least quantity of effluents and can be reused or discharged with minimal treatment. Cleaning wastewater originates from the cleaning process of equipment and may lead to malfunctioning of equipment and other operational errors, while sanitary wastewater is directly piped to sewage works. Proper treatment of these waste waters is a necessary task.

Dairy wastewater generally contains milk, other byproducts, cleaning products and various preservatives and additives. Bovine milk is composed of 87% water, 4.7%lactose, 4% lipids, 3.5% protein and 0.8% ash (Bylund 1995). The fat contents of milk range between 3-5% where 98% of which consists of triglycerides. Other important fat contents include sterols (0.2-0.5%) and phospholipids (0.5-1%) (Jensen *et al.* 1995). Milk also contains trace amount of other organic compounds, such as terpenes, depending on feeding of the lactating animal (De Noni *et al.* 2008). The composition of dairy wastewater also varies with the composition of milk and type of product to be manufactured. Condensate collected from the evaporation of milk is considered as the cleanest wastewater and contains only some volatile organic compounds (Verheyen *et al.* 2009). Dairy wastewater can also be applied to the land in some parts of the world. Some industries add combination of potassium hydroxide and sodium hydroxide to lower down the concentration of sodium in the effluent to be applied on the land. As dairy industries are producing a variety of products by using various chemicals so dairy processing wastewater is highly variable in terms of composition.

Whey as a Dairy Industry Waste

Whey is the milk serum obtained after removing fat and casein. On the basis of methods of coagulations, whey can be divided into two broad categories i.e., sweet and sour whey. Sweet whey is produced as a result of enzymatic coagulation of milk. Mostly proteolytic (chymosine and pepsine) and microbial enzymes produced from *Mucor miehei* and *Mucor pusillus* are involved in this process. It is produced during the manufacturing of different types of cheeses such as Italian, cheddar and Swiss cheese. Sour whey is obtained by acidic coagulation of milk mainly due to lactic acid fermentation. It is produced as a result of soft cheese production. Whey generally contains approximately half percentage of the solid contents (6.5%) of the whole milk and also contains significant quantities of water soluble vitamins and minerals. The solids of the whey include 68-72% lactose, 12-13% proteins 8-9% minerals and traces of lactic acid and fat (Marisa 2004). Whey was previously thought to be a waste product but now it is commercially available in the market as protein concentrate, lactose, lacto peroxidase and lactoferrin (Prasad *et al.* 2004). It has also been used extensively as energy drink for athletes (Brodovich 2006).

The composition of whey depends on milk composition which varies with species of animal, breeding practices, climate, lactation stage, feeding and differences in the individual animal (Polat 2009). About 90-95% of the milk volume is cheese-whey that keeps approximately 55% of the milk nutrients. Whey usually contains 4.5-5.0% lactose, 0.6-0.8% proteins, 0.4-0.5% fats and 8.0-10% mineral salts. Commonly present salts include NaCl, KCl and calcium salts. Additionally, cheese-whey also contains lactic acid, citric acid, urea, uric acid and vitamins of B group. Acid or sour whey also keeps significant quantities of salt but lower protein substances than sweet whey. Due to high salinity, it is quite difficult to process and dispose-off salty whey as compared to sweet whey (Blaschek *et al.* 2007).

The annual production of whey is about 160 million tons around the world and a continuous increase of 1-2% per annum is recorded (Guimaraes *et al.* 2010). It is jaggedly estimated that about half of the world whey production is discarded as waste without treatment. It causes thoughtful environmental and health hitches because of the higher volume of organic matter. Furthermore, it also carries significant amount of lactose which is mainly liable for the high biological oxygen demand (BOD) and chemical oxygen demand (COD) exhibition in the naturally polluted media (Mawson 1994). Whey has 30-50g/L BOD 60-80g/L COD due to its high organic contents which pollute the streams and rivers. Domingues *et al.* (1999) pointed out that protein recovery reduced COD of whey up to 10 g/L. Likewise, volatilization of cheese-whey has also been the focus of investigation since decades (Turkmenoglu 2006; Guimaraes *et al.* 2010). Additionally, reduction of BOD (>75%) has been reported with the concomitant production of ethanol, biogas or single cell proteins (Siso 1996), thus it can be posited that about half of the total whey production has now been utilized for different purposes.

TREATMENT OF DAIRY WASTE

Waste Water Treatment

Waste water is treated simply by removing the solid impurities by mechanical sedimentation. Sometimes, biological treatment is also applied with this process to increase the efficiency and for more effective decomposition of organic compounds present in the waste water. Additionally, chemical treatment of waste water can also be merged with these treatments for removing phosphorus compounds and the process is known as post-precipitation. Furthermore, mechanical treatment can also be combined with chemical precipitation for enhancing the efficiency of process. In this process, most of the sewage is treated in a single step resulting in about 90% reduction of phosphorus. As a result, the biological stage requires less basin volume and energy input. The choice of treatment methods depends upon the required degree of pollutant reduction. There are four possible systems.

Post, Pre, Direct and Simultaneous Precipitation

Post-precipitation corresponds to a traditional three-stage method with mechanical-biological-mechanical treatment. It is an effective but expensive technique. While pre-precipitation concerns two-stage process in which a combination of chemical and mechanical treatments is used resulting in a significant reduction of phosphorus and about 70% reduction of BOD. This step is quite effective in reducing the energy requirements and basin volume during biological treatment.

Regarding direct precipitation, it is a single-stage method in which chemical and mechanical treatments are applied simultaneously as a single step with no subsequent biological treatment. During simultaneous precipitation treatment of waste water, mechanical treatment is combined with biological-chemical treatment as a simultaneous process. This method is considered as a quite economical process resulting in significant BOD and phosphorus reduction without additional basin capacity.

Mechanical, Chemical and Biological Treatments

The mechanical treatment of dairy waste water involves the use of sedimentation basins, strainer grid and sand trap. The grid is used for trapping the plastic particles, coarse solid materials and food particles etc. The sand trap is used for the separation of remaining coarse particles from the system. The operation of sand trap basin allows sand and heavy solid particles to settle down and keeps fat and other lighter impurities floating on the surface. The settled material is pumped out and floating scum is eliminated by using the scrapers. Additionally, air is blown into the sand trap with pressure to keep the trapped and smaller particles in the suspension which prevents putrefaction and off-odor (Bylund 1995).

The prime objective of chemical treatment of waste water is to remove the phosphorus-containing compounds from the waste. Municipal sewage system generally collects 2.5-4 g of phosphorus per person on daily basis as phosphates. Around 30% of the phosphate contents come from the detergents whereas remaining 70% portion of the phosphorus-containing compounds is obtained from food residues and human excreta. Chemical precipitation method is more effective as compared to the biological treatment as it removes almost 100% phosphorus during treatment while biological treatment only reduces 20-30% of the phosphorus present in the wastewater.

The precipitation is carried out in flocculation tanks, where water is mixed vigorously with flocculants by agitators resulting in the initial accumulation of phosphates followed by aggregation. After aggregation, the larger flocks are settled down in the pre-sedimentation basins, while the finer effluents overflow for further biological treatment. Pre-sedimentation is regarded as the last step of chemical and mechanical treatments of wastewater. The water is then flown through the basins and the finer particles settle down gradually to the bottom of the basins as primary sludge. These basins are also equipped with the devices for continuous scraping of the sediments into a sump that carry off water from the clarified surface layer (Bylund 1995). During the biological treatment of wastewater, the organic compounds, present in the wastewater, are broken down with the help of microorganisms. The micro-organisms perform biological digestion in aerobic environment and must need oxygen which is supplied as air blown into the basins. As a result of microbial activities, the active sludge is produced.

The generated sludge is then removed from the system through settling and transferred to the post-sedimentation basins followed by recycling in the aeration basins for further biological breakdown. Finally, the excess sludge is removed from the system and the clarified effluent is discharged. Sometimes, biological filters can also be used as an alternative to the aeration basins, which contain pieces of plastic and stones. A rotating distributor is used to sprinkle the water over the biological filter and water is then moved down to the filter bed. Air circulation is provided for microbial activities. A "layer of microorganisms is built up on the surface of stones due to microbial activities (Bylund 1995).

Aerobic and Anaerobic Treatment of Dairy Wastewater

The dairy industry applies aerobic or anaerobic treatment for the treatment of the effluent. Aerobic techniques need a source of power to provide the fresh and clean air needed to incorporate the natural matter in the wastewater and hence are more suitable to decrease the average strength of the wastewaters, since the greater the organic matter the higher the fresh air required and the greater the costs. Anaerobic techniques have been employed for their efficiency to treat the wastes of great strength and utilization of the methane gas.

In aerobic treatment techniques, bacteria, in the existence of O_2, turn the natural elements of the waste material to CO_2, H_2O and microbial biomass. The aerobic treatments have the ability to cause smells if managed wrongly. The market globally has utilized many types of aerobic techniques including trickling filtration, spinning scientific contactors and different types of automatically oxygenated lagoon techniques. In New Zealand, only prolonged aeration triggered debris techniques are used. Common treatment factors for an activated sludge dealing with dairy wastewater are 94% Chemical oxygen demand, 99% biological oxygen demand5 70% TKN and 50% total phosphorus content removal (Dyatlova *et al.* 2008).

In an anaerobic digestion chamber, anaerobic microbes, performing in the lack of fresh air, turn the natural elements in the wastewater to CH_4, CO_2 and H_2O. Organic types of nitrogen are transformed to the ammonium nitrogen. Anaerobic digestive function may be conducted in low speed lagoon techniques or in high speed reactors. The more latest anaerobic digesters which have been set up in the dairy industry have been high-speed digesters, usually with two stages to acquire better management of the anaerobic techniques (Dyatlova *et al.* 2008).

The merits of anaerobic digestive function are:

- Production of a valuable by-product (methane), that can be retrieved and used as a fuel
- Remove significant amounts of BOD5 and COD without the input of technical power for aeration
- Produce lower debris than aerobic techniques

Anaerobic treatment is further categorized into two processes:

- Single-phase anaerobic treatment
- Two-phase anaerobic treatment

Anaerobic treatment processes are usually preferred over aerobic processes due to their high efficiency and benefits for treating wastewater (Rajeshwari *et al.* 2000). Anaerobic filters have been widely used around the globe for the treatment of dairy wastewater. These are mostly suitable for the wastewater containing lower concentration of suspended solids. Plastic anaerobic filters generally reduce COD up to 78 to 92% during a hydraulic retention time (HRT) of 4 days (Viraraghavan and Kikkeri 1991a). Additionally, these anaerobic filters are effective over a wide range of temperature between 21 to 30°C (Viraraghavan and Kikkeri 1991b).

Generally, two types of anaerobic filter reactors namely; porous and non-porous filter media are used for the treatment of wastewater in ice-cream and milk bottling industries. Studies reveal that filters porous support media showed better results in comparison with filters having non-porous media (Anderson *et al.* 1994). Another investigation concluded that use of anaerobic filter reactor for the treatment of wastewater in an ice-cream production plant reduced COD up to 70% at an average load of 5.5 kg COD/(m^3 day) (Monroy, *et al.* 1994). In another work, dilute dairy wastewater was treated by employing upflow anaerobic filter reactor (UAF) at an organic loading rate (OLR) range of 0.117 and 1.303 g volatile solids per day and the results confirmed a satisfactory reduction in the COD during the treatment.

Two-phase anaerobic treatment system is utilized for the wastewater containing higher concentration of organic effluents (Demirel *et al.* 2005). This system involves stabilization of waste by pre-acidification followed by anaerobic filtration of the waste material. The efficiency of acid phase reactor is of great importance in two-phase anaerobic waste stabilization since it provides a suitable substrate for the subsequent methane phase reactor (Alexiou *et al.* 1994). A number of investigations have been conducted to evaluate the performance of two-phase anaerobic treatment of dairy wastewater. Initially, lactose was degraded by acidogenesis in laboratory-scale studies (Huang and Pinder 1995). Additionally, different other types of dairy wastes were also treated by using mesophilic two-phase system (Strydom *et al.* 1997). Results of the studies showed 97% COD reduction at an OLR of 2.82 kg COD/(m^3 day) for cheese manufacturing effluents, whereas for fresh milk effluents 94% COD removal was achieved at an OLR of 2.44 kg COD/(m^3 day). Performance levels of both types of anaerobic treatments of dairy wastes are presented in Table 1.

Submerged Membrane Filtration Treatment

Wastewater treatment systems are usually designed with a purpose to alleviate the impurities to meet discharge regulations. Different physico-chemical and biological methods have been employed for the treatment of dairy industry waste (Vidal *et al.*

Table 1. Comparison of performance levels for single-phase and two-phase anaerobic treatment of dairy wastewaters

Process	Waste type	Reactor type	HRT	Loading (kg COD/m^3 day)	Temperature (°C)	Methane yield (m3/ kg COD)	Removal (%)
Single-phase anaerobic treatment	Cheese whey	Downflow fixed film	4.9 (day)	13	12.5-30	0.28-0.34	75-76 (COD)
	Ice-cream	Anaerobic filter	0.5 (day)	5.5	NA	0.32-0.34	75 (COD)
	Raw milk	Anaerobic filter	NA	5-6	NA	-	90 (COD)
	Synthetic effluents	Hybrid	1.7-4.1 (day)	0.82-6.11	35	0.35	90-97 (COD)
Two-phase anaerobic treatment	Dilute milk waste	CSTR+upflow filter	4.4 (day)	9.4	35	-	92 (COD)
	Wastewater of milk bottling plant	CSTR+upflow filter	2 (day)	NA	35	-	90 (COD)
	Skimmed milk	CSTR+upflow filter	2 (day)	NA	20	-	95.5 (COD)
	Wastewater of milk and cream bottling plant	CSTR+upflow filter	2 (day)	NA	33-36	-	90 (COD)

HTR= Hydraulic retention time

CSTR= Continuously stirred tank reactors

2000). However, traditional physic-chemical methods are not considered effective for the removal of contaminants from dairy waste water due to their poor efficiency and high cost (Sharma 2014). Contrarily, biological systems are generally considered a better alternative because of their high efficiency (Bangsbo and Hansen 1985). The most commonly employed biological treatments for dairy wastewater treatment include anaerobic treatment, treatment in ponds and activated sludge plants (Demirel *et al.* 2005). In recent years, several new, membrane-based techniques like microfiltration, nano-filtration, ultrafiltration and reverse osmosis have been investigated for the treatment of wastewater. These techniques have proven through various studies as energy efficient in comparison with the conventional methods (Mavrov *et al.* 2001). Mechanistic approach involved behind the membrane filtration technology is selective permeability of components in a liquid mixture through a membrane barrier. Studies also conclude that deployment of reverse osmosis or nano-filtration for dairy wastewater treatment is quite helpful to concentrate milk components present in the wastewater stream (Jelen 1991). The inferences of the investigation revealed that single filtration operation significantly concentrated the milk components in the retentate. Additionally, integration of biological and membrane filtration methods has also shown better filtration quality for the treatment of dairy wastewater. The average removal efficiencies of using submerged membrane microfiltration system for BOD, COD and total dissolved solid (TDS) were 98.8%, 92.5%, 96.7% respectively. (Al-Shammari *et al.* 2015).

Whey Management/Treatment

Whey is a major byproduct of dairy industry in Pakistan as well as worldwide. It is a milk serum obtained after coagulation of milk in cheese making and extraction during butter making process. It usually contains lactose, proteins, minerals and some vitamins. Whey disposal is a very costly and time-consuming process its direct disposal into sewerage water causes water pollution. It is a cheaper organic nitrogen and carbon source and can be utilized for the production of value-added products which can be used in food and beverage, pharmaceutical and textile industry. A survey concludes that approximately 50% of the whey production is treated for further utilization in various food products. However, this percentage can be increased by research efforts in whey utilization (Guimaraes *et al.* 2010). New technologies have been developed for treatment of whey and waste water. According to a general estimate, about 45% of the total whey is used by the food industry as liquid, 30% in the form of whey powder, 15% as byproducts and the rest as whey protein concentrates (Kotoupas *et al.* 2007).

There are three different ways to utilize whey. The first one is direct use of whey in which includes the incorporation of whey in animal feed and other food products.

Second category is direct stabilization in which whey is physically or chemically treated to make it more stable against microbial dilapidation. It includes recovery of proteins by ultrafiltration, heat denaturation and concentration of whey protein by evaporation or reverse osmosis. Third technique is the conversion in which lactose is converted into other products by biotransformation or chemical reactions (Mawson 1994). Raw whey can be treated in several ways. Some important and modern techniques of whey processing are discussed in the following lines.

Regarding condensers for which the first step of whey processing is evaporation to condense total volume. Raw whey contains 6% dissolved solids which are to be raised to 40-50%. After that dry powder is produced by spray drying. Different types of condensers are being used for condensation purpose. One type of condenser involves different evaporative stages to boil the whey below 212ºF temperature by creating negative pressure. This process is energy efficient because in these types of condensers one effect acts as a source of heat for next effect by providing its latent heat of vaporization. The whey proteins being vaporized enters the pipe enclosed by the fumes in the external covering shells.

Another technique of evaporation is known as mechanical vapor compression evaporator that involves reprocessing of fluid in pipes enclosed by warm vapor. Some of the fluid evaporates and is pulled into the high-power air compressor which enhances pressure and causes the hot fumes into contact with the evaporation pipes. The fumes, being warmer than the whey proteins fluid in the pipes, provide their latent heat of vaporization thus making the procedure more power effective. Both moisture build-up or condensation procedures produce a wastewater known as condensate of whey proteins. The procedure also produces wastewater from cleaning the handling and processing devices.

The endemic use of ultrafiltration has started out up new industry for whey proteins items. The ultrafiltration is used for the separation of protein compounds from the lactose and other elements in the whey. High pressure is used to pass the smaller elements through the membrane and large elements remains in the concentrated flow. This protein-flow is compacted and dried to process the whey protein. Edible proteins have been widely used as protein supplements and food additives. The lactose sugar portion (permeate) can also be compacted, dried and used in baby meal preparation, as sugar in bakery products and in medicines as a service provider in pharmaceutical industry. Moreover, huge amounts of permeates are distributed on crop farms as potassium, nitrogen and phosphorus manure. Technology is also available to produce items such as ethanol, methane gas and yeast cells.

The Reverse osmosis process involves the selection of specific molecular membrane to filter water molecules thus concentrating the whey proteins solids. Solid contents of whey are increased from 6% to 12% by reverse osmosis to prevent the transportation expenses and evaporator potential.

Biological treatment, relating to the use of anaerobic digestion systems, is an effective procedure to decrease the pollutant content if cheese manufacturing industries have convenience problems and cannot manage the great investment expenses for whey protein valorization. Because of the great organic matter of whey proteins, anaerobic digestive function comprises an excellent remedy (Erguder *et al.* 2001). The anaerobic digestion procedure includes the deterioration and stabilization of organic matters by bacteria under anaerobic conditions and results in biogas (methane) and biomass production (Najafpour *et al.* 2008).

In particular, with respect to whey proteins, this complicated procedure includes three subsequent steps: breakdown of proteins and lactose, fermentation and methanogenesis (Audic *et al.* 2003). Several bacterial strains are used in this process. Approximately 90% of the hydrolyzed matter is turned into methane gas in this process. It is estimated that 1 L of whey proteins generates 45 L of biogas having 55% methane and the predicted oxygen demand elimination is 80%. For each liter of whey proteins, 20 liters of methane can be created. However, use of anaerobic digestion is not usually practiced in the dairy industries due to its low rate of reaction and the comparative procedure uncertainty in traditional reactors (Gannoun *et al.* 2008).

In common, it is difficult to evaluate techniques managed in various labs, not only because of the possible variations in the anaerobic waste features (Gannoun *et al.* 2008), but also due to the variations in working aspects such as heat range, pH, organic load rate, total solids and volatile fat content, poisoning, gas preservation time, type of inoculums and composition of the effluent.

Since anaerobic digestion is a biochemical procedure motivated by various bacteria which convert complicated organic matter into simple substances methane, any stress or interference on the system may results a change in species and population levels, which is gradually shown in the reactor efficiency. Therefore, the organic material involved as inoculum in the fermentative organic substrate is one of the major elements affecting anaerobic digestion of organic material. High microbial load inoculum material needs to be involved in the combination of biomass when starting up the anaerobic digestion. The inoculum has significant effects on biogas manufacturing (Mateescu and Constantinescu 2011). The selection of the appropriate inoculum is very essential and relies on the structure of the substrate (C/N/P, alkanity). For example, Gannoun *et al.* (2008) revealed that for anaerobic digestive function of pretreatment of milk products whey proteins with the best possible composition for the anaerobic digestive function program, the inoculum can be acquired from an effective biogas digester of fruits and vegetable waste therapy (Bouallagui *et al.*, 2013).

Depending upon different kinds of design requirements, anaerobic digestion is classified into different kinds according to feeding technique (batch, semi-batch or continuous reactors), the range (thermophilic, mesophilic and psychrophilic reactors), the solid material (high or low strong reactors), the complexness (single stage or two (multi) stage reactors), the shape of the reactor (horizontal and vertical), the way microbes are handled in the reactor (fixed film, suspended growth and hybrid), the moisture in the substrate (wet or dry digestion) (Qi 2008).

Sludge Treatment

With regards to both investment and working costs, sludge treatment and disposal can be as expensive as the rest of the effluent treatment. Improving attention of how waste disposal impacts our atmosphere has decreased the options available for disposal and improved the costs, and this pattern is likely to continue.

The sludge from the various stages of treatment is collected in thickening tanks to which chemicals are added to facilitate further aggregation of the solid particles. To further break down organic matter and to reduce bad-smelling substances, the sludge is eventually pumped into a digester where the organic substances are broken down under anaerobic conditions into carbon dioxide and methane and very small amounts of hydrogen gas, ammonia and hydrogen sulphide. Carbon dioxide and methane are the main components of digester gas, which can be utilized as fuel for heating. Digester sludge is a homogeneous, practically odorless, dark-colored substance which still has high moisture content varying in the range of 94-97%. It is therefore dewatered, most effectively in a decanter centrifuge, which discharges a solid phase of about one-eighth of the original volume. The dewatered sludge can then be utilized as fertilizer or landfill, or simply deposited as waste (Bylund 1995).

Different sludge treatment methods are used either to decrease the amount of sludge for disposal, or to turn it into a simple form either appropriate for re-use (e.g. area spreading) or for dump. Sludge thickening is used in case of additional biological waste sludge and main solids. Primary solids are lightweight and settle easily without the use of chemical treatment and the H_2O in these solids is not highly organized. Chemical substances or additives are needed for the secondary treatment of the sludge in order to improve dewatering. Primary and secondary sludge are mixed together in order to improve the dewatering procedure. This decreases the percentage of entrained H_2O.

A traditional sludge thickener usually makes the sludge thick up to 4-8% dry solids. For many set ups this is adequate to decrease the debris volume to a level that is financially feasible off-site convenience. For bigger sites, the thickening procedure is basic for further dewatering process. Sludge dewatering improves the dry solid sludge material, generating waste in solid form. This waste consists of up to 20-50% dry solids, which considerably decreases disposal expenses. Dewatering is usually done by using chemical additive substance for the sludge and then by employing filtration or centrifugation. Different filtration methods can be used for the treatment of dairy sludge, for intense:

- Using filter press that can be manually intense, and filter cake contains almost 40% dry solids
- Deploying belt press that generates a filter cake containing upto 35% dry solids and usually have high chemical expenses
- Use of screw press designed for the waste material with a higher percentage of main tests, and can generate a filter cake containing 25-30% dry solids

The dry solids centrifuging which is a continuous procedure generate a cake containing up to 40% dry solids for certain debris. Smell problem is very little because it is a closed procedure.

UTILIZATION OF DAIRY INDUSTRY WASTES

Waste-Water Utilization

Dairy industry wastewater is used for irrigation because it is rich in nutrients and can provide valuable phosphorus and nitrogen to land (add reference). However, it can affect land negatively and can change soil structure (soil porosity) causing high infiltration rate, pollution of water sources and groundwater, and suppress plant growth. In wastewater irrigation, wastewater salt, and Ca, Mg, P, N minerals as well as the land under irrigation is all of worth consideration (Overcash and Pal 1979). In badly handled systems excess nutritional value, especially nitrates, may leach through the ground and can contaminate groundwater. Both N and P may be released offsite in surface area run-off and speed up eutrophication of receiving sites or ponds. Treated wastewater, however, contains less nutritional value leading to lower chances of contamination of ground and surface water, but still it has a threat associated with the relatively high salt content.

The dairy effluents provide a phosphorus to land without causing negative ecological effects and reduces the requirement for phosphorus manure. Phosphorus is lost from irrigation sites due to break down and draining of contaminants through ground pores (Leinwebe *et al.* 2002).

Carbon have very important role in terms of improving the soil structure and enhancing microbial activity of the ground. Nitrogen is one of the three main nutrients required by plants. Most of the nitrogen is present in the soil in form of organic matter. Nitrogen material in wastewater is valuable nutrient resource for soil and reduces the need for nitrogen fertilizer. Some nitrogen is available in free form for direct use by plants, but natural nitrogen must be mineralized by bacteria to ammonium by ammonification or nitrate by nitrification so that plants can utilize it. Excessive nitrogen in the effluent that is higher than the needs of the plants may be released through surface run-off or draining.

The carbon and nitrogen proportion are an essential attribute of dairy plant wastewater when it is used for land irrigation. High levels of quickly degradable carbon can increase the microbial action. This higher microbial action can increase the competition between plants to consume nitrogen which results decreased amount of available nitrogen to vegetation. Thus, the application of the untreated dairy wastewater to field results in reduction of plant growth (Ghani *et al.* 2005).

Whey Utilization

Whey protein represents 15-22% of the total milk proteins. There are five major protein fractions with described molecular weights, i.e., immunoglobulins (Igs), bovine serum albumin (BSA), α-Lactalbumin (α-La), β-Lactoglobulin (β-Lg) and proteose peptones (Merina *et al.* 2001). The pure whey protein concentrate (WPC) normally contains 50-85% protein on dry basis, while whey protein isolate (WPI) contains 90-98% protein and very small portion of fat and lactose. Whey protein is top quality protein having 3.4 protein efficiency ratio (PER), equals to albumin and 2.8 greater than casein (Siso 1996).

Half of the whey protein produced is treated and turned into different food and feed items. About 50% of the whey protein is directly used as liquid, 30% in powdered form, 15% as lactose and the rest (5%) is used as pure whey protein concentrate. Due to its features, cheese-whey can be used as component of many items. Cheese whey protein is used as an ingredient of new whey protein items (Turkmenoglu 2006). Liquid whey protein can be used for either for the production of bio-fertilizer or as a feed for the farm animals. Powder whey protein is used in animal feed and also used in human food as sweetening agent. Another very important use of whey protein is using whey as a protein source. It can be turned into pure WPC and used as food additive and food component. Nowadays ultrafiltration and of whey protein are standard functions in the dairy market (Pouliot 2008) they allow their recovery without significant loss and their efficient qualities having low salt, making it apt for people to use.

Whey protein hydrolyzates are used as protein supplements for baby formulas, sportsmen and muscle builders. These proteins are more quickly consumed but the hydrolysis level should be properly managed to prevent the development of the nasty proteins (Mann 2000).

The determination of the physico-chemical and efficient qualities of pure whey protein elements and the development of protein fractionation, mostly membrane fractionation, are on the basis for the development of new applications. β-Lg and α-La fractions are now created at industrial level (Smithers, 2008). Whey protein hydrolyzates fractionation initiates the development of bioactive proteins (e.g., angiotensin transforming enzymes and other antihypertensive proteins as pointed by several previous investigations ((Fitzgerald and Murray 2006; Welderufael and Jauregui 2010; Illanes 2011).

Whey protein is also used for the development of antianaemics and iron propionates. Permeate with a higher lactose material is produced during the production of whey protein concentrate as a by-product. It is essential to consider that protein restoration does not fix polluting environment problems (Mawson 1994). Moreover, lactose can be retrieved from cheese whey. Lactose is also used as complement in baby meals; cow's milk items contain 30% less lactose as compared to human milk. It is also used as a building agent in confectionary products due to its low sweet taste than mono or disaccharides. Whey is also used in meat and bakery items. The limited use of lactose in food is due to its low solubility and lactose intolerance. Lactose is inert, non-hygroscopic, and available with great hygiene and has good executed qualities therefor it is used as excipient for tablets in pharmaceutics (Fox 2009).

The manufacturing of methane by fermentation of whey protein is a procedure consisting of three steps: (i). hydrolysis of lactose and protein, (ii) fermentation (iii) methanogenesis. Methane production is a very complex procedure it includes several microbial varieties.

Methane manufacturing through cheese whey protein fermentation symbolizes an important resource of energy as fuel to produce electricity. Hypothetically, 1 kilo gram of lactose results in 0.75 m^3 of biogas production containing roughly 50% vol/vol methane. For a deproteinated whey protein the theoretical generate is roughly 20.7 m3 methane/m^3 (equal to 18.6 L of energy oil) (Kosaric and Asher 1985).

Lactose and whey protein is used as a source for the development of microbial biomass. Several industries produce microbial biomass from dairy waste have been revealed. Biomass created from lactose or whey protein (Champagne *et al.* 1990) is mostly used for manufacturing of baker's yeast and as animal feed complement. These processes are also of most interest in treatment of dairy wastewater: over 80% BOD is reduced in wastewater treatment with the production of dry yeast of half of the weight of the milk sugar used (Gonzalez 1996).

Whey is also used for the development of alcohols mainly ethyl alcohol (Castillo 1990). High ethyl alcohol (about 80%) can be achieved from *Kluyveromyces fragilis* stresses. Ethyl alcohol can be used as a source of energy (fuel) or to generate acetic acid. Many distilleries generate ethyl alcohol from pure whey protein in Ireland (Europe), the America and New Zealand about half of pure whey protein manufacturing is converted into ethanol (Gonzalez 1996). Carbon and nitrogen are substrates for microbes in whey protein (vitamins and fats) required for living microbial growth. Breakdown of lactose is caused by β-galactosidase (from a micro-organism). Combination of sugar and galactose is used as carbon source by *Saccharomyces cerevisiae*. Additionally, a combination of acetone, butanol and ethyl alcohol can be acquired by fermentation of pure whey protein (Kilara and Patel, 1992) but the commercial prospective of the technique seems low presently. By fermentation of cheese pure whey protein 2-3 butanediol is produced it is used as alternative source of energy in the chemical industry (Sienkiewicz and Riedel 1990).

Acetic, propionic, lactic, lactobionic, citric, gluconic and itaconic acids can be acquired from whey fermentation (Hwang and Hansen 1998). Most of them find application in chemical industry. Lactic acid is used to generate propylene oxide, eco-friendly polylactic acid polymers, plastic fibers or propylene glycerin, bearing the lactic acid market of primary significance (Lunt 1998). Recent interest in eco-friendly plastic could lead to enhanced requirement for use of lactic acid in plastic manufacturing. Acetic acid is also used as a raw material in the chemical industry to generate calcium magnesium acetate that finds its application as road or airport terminal de-icer (Yang and Silva 1995). Propionic acid is used for the development of food preservatives, weed killers and intermediate chemicals.

CONCLUSION

The dairy industry is one of the most polluting sectors, not only with regards to the number of effluents produced, but also with regards to its features as well. Dairy effluent contains dissolvable organics, suspended solids and trace organics. All these elements play a role mostly towards their high BOD and COD requirement. Dairy effluent treatment and usage is of major concern these days. Its high organic matter content along with the high amounts of effluent created and limited treatments make dairy industry effluent a serious ecological problem. However, the prospective manufacturing of biogas (methane), hydrogen or other valuable products with a multiple high COD reduction through appropriate treatment shows that milk market waste must be considered as an energy source rather than a pollutant.

Pure whey protein is a by-product that has been typically considered as remains of dairy products manufacturing. Only a few companies have the capacity to apply different technological innovation for the efficient restoration and improving of whey. Fermentation process of lactose and whey provides a variety of valuable products such as methane, alcohols, natural chemicals, baker's yeast and vitamins.

In addition to in-plant recycle, dairy industry wastewaters can be used to irrigate field or public grounds. However, there are potential risks associated with the use of dairy waste water. More research is needed to develop some new, rapid and safe methods to treat the dairy waste in a much effective way.

CONFLICTS OF INTEREST

None

REFERENCES

Ahmad, T., Aadil, R. M., Ahmed, H., Rahman, U. R., Soares, B. C. V., Souza, S. L. Q., ... Cruz, A. G. (2019). Treatment and utilization of dairy industrial waste: A review. *Trends in Food Science & Technology*, *88*, 361–372. doi:10.1016/j.tifs.2019.04.003

Al-Shammari, S. B., Bou-Hamad, S., Al-Saffar, A., Salman, M., & Al-Saira, A. (2015). Treatment of dairy wastewater effluent using submerged membrane microfiltration system. *Journal of Environmental Science & Engineering*, *4*, 107–118.

Alexiou, I. E., Anderson, G. K., & Evison, L. M. (1994). Design of pre-acidification reactors for the anaerobic treatment of industrial wastewaters. *Water Science and Technology*, *29*(9), 199–204. doi:10.2166/wst.1994.0479

Anderson, G. K., Kasapgil, B., & Ince, O. (1994). Comparison of porous and nonporous media in upflow anaerobic filters when treating dairy wastewaters. *Water Research*, *28*(7), 1619–1624. doi:10.1016/0043-1354(94)90229-1

Bangsbo, D., & Hansen, I. (1985). Treatment of dairy wastewater in the developing countries-The Danish experience. *India Environment*, *8*, 2–10.

Blaschek, K. M., Wendorff, W. L., & Rankin, S. A. (2007). Survey of salty and sweet whey composition from various cheese plants in Wisconsin. *Journal of Dairy Science*, *90*(4), 2029–2034. doi:10.3168/jds.2006-770 PMID:17369245

Bouallagui, H., Ben, C. R., Marouani, L., & Hamdi, M. (2003). Mesophilic biogas production from fruit and vegetable waste in a tubular digester. *Bioresource Technology*, *86*(1), 85–89. doi:10.1016/S0960-8524(02)00097-4 PMID:12421014

Brodovich, H. O. (2006). *Energy Drinks, Whey Protein, and More: Dietary Supplements and Teen Athletes*. The Hospital for Sick Children.

Bylund, G. (1995). *Dairy Processing Handbook*. Lund: Tetra Pak Processing Systems.

Castillo, F. G. (1990). Lactose metabolism by yeast. In H. Verachtert & R. De-Mot (Eds.), *Yeast Biotechnology and Biocatalysis* (pp. 297–320). New York: Marcel Dekker.

Champagne, C. P., Gioulet, J., & Lachance, R. A. (1990). Production of baker's yeast in cheese whey ultrafiltrate. *Applied and Environmental Microbiology*, *56*, 425–430. PMID:16348117

Chong, M. L., Sabaratnam, V., Shirai, Y., & Hassan, M. A. (2009). Biohydrogen production from biomass and industrial wastes by dark fermentation. *International Journal of Hydrogen Energy*, *34*(8), 3277–3287. doi:10.1016/j.ijhydene.2009.02.010

De Noni, I., & Battelli, G. (2008). Terpenes and fatty acid profiles of milk fat and "bitto" cheese as affected by transhumance of cows on different mountain pastures. *Food Chemistry*, *109*(2), 299–309. doi:10.1016/j.foodchem.2007.12.033 PMID:26003351

Demirel, B., Yenigun, O., & Onay, T. T. (2005). Anaerobic treatment of dairy wastewaters: A review. *Process Biochemistry*, *40*(8), 2583–2595. doi:10.1016/j.procbio.2004.12.015

Domingues, L., Dantas, M. M., Lima, N., & Teixeira, J. A. (1999). Continuous ethanol fermentation of lactose by a recombinant flocculating *Saccharomyces cerevisiae* strain. *Biotechnology and Bioengineering*, *64*(6), 692–697. doi:10.1002/(SICI)1097-0290(19990920)64:6<692::AID-BIT8>3.0.CO;2-J PMID:10417218

Dyatlova, G., Pevnev, S., & Fedorovskaya, T. (2008). Treatment of milk plants wastewater. *Water Supply and Sanitary Technology*, *2*, 12–15.

Erguder, T. H., Tezel, U., Guven, E., & Demirer, G. N. (2001). Anaerobic biotransformation and methane generation potential of cheese whey in batch and UASB reactors. *Waste Management (New York, N.Y.)*, *21*(7), 643–650. doi:10.1016/S0956-053X(00)00114-8 PMID:11530920

Fitzgerald, R. J., & Murray, B. A. (2006). Bioactive peptides and lactic fermentations. *International Journal of Dairy Technology*, *59*(2), 118–125. doi:10.1111/j.1471-0307.2006.00250.x

Fox, P. F. (2009). Lactose: Chemistry and properties. In Advanced Dairy Chemistry (3rd ed.). New York: Springer.

Gannoun, H., Khelifi, E., Bouallagui, H., Touhami, Y., & Hamdi, M. (2008). Ecological clarification of cheese whey prior to anaerobic digestion in up flow anaerobic filter. *Bioresource Technology*, *99*(14), 6105–6111. doi:10.1016/j.biortech.2007.12.037 PMID:18242084

Ghani, A., Dexter, M., Sarathchandra, U., & Waller, J. (2005). Effects of dairy factory effluent application on nutrient transformation in soil. *New Zealand Journal of Agricultural Research*, *48*(2), 241–253. doi:10.1080/00288233.2005.9513653

Gonzalez, S. M. I. (1996). The biotechnological utilization of cheese whey: A review. *Bioresource Technology*, *57*(1), 1–11. doi:10.1016/0960-8524(96)00036-3

Guimaraes, P. M. R., Teixeira, J. A., & Domingues, L. (2010). Fermentation of lactose to bio-ethanol by yeasts as part of integrated solutions for the valorization of cheese whey. *Biotechnology Advances*, *28*(3), 375–384. doi:10.1016/j.biotechadv.2010.02.002 PMID:20153415

Huang, J., & Pinder, K. L. (1995). Effects of calcium on development of anaerobic acidogenic biofilms. *Biotechnology and Bioengineering*, *45*(3), 212–218. doi:10.1002/bit.260450305 PMID:18623140

Hwang, S., & Hansen, C. L. (1998). Characterization and bio-production of short chain organic acids from mixed dairy-processing wastewater. *Transactions of the ASAE. American Society of Agricultural Engineers*, *41*, 795–802. doi:10.13031/2013.17217

Illanes, A. (2011). Whey upgrading by enzyme bio-catalysis. *Electronic Journal of Biotechnology*, *14*(6), 1–28. doi:10.2225/vol14-issue6-fulltext-11

Jai, P. K., Vimal, C. S., & Indra, D. M. (2011). An overview of various technologies for the treatment of dairy wastewaters. *Critical Reviews in Food Science and Nutrition*, *51*(5), 442–452. doi:10.1080/10408391003663879 PMID:21491269

Jelen, P. (1991). *Pressure-driven membrane processes: Principles and definitions*. Brussels, Belgium: International Dairy Federation.

Jensen, R. G., Ferris, A. M., & Lammi-Keefe, C. J. (1991). The composition of milk fat. *Journal of Dairy Science*, *74*(9), 3228–3243. doi:10.3168/jds.S0022-0302(91)78509-3 PMID:1779072

Kilara, A., & Patel, M. T. (1992). Whey and lactose fermentation. In J. G. Zadow (Ed.), *Whey and Lactose Processing* (pp. 409–448). London, UK: Elsevier Applied Science. doi:10.1007/978-94-011-2894-0_11

Kosaric, N., & Asher, Y. J. (1985). The utilization of cheese whey and its components. *Advances in Biochemical Engineering*, *35*, 25–60.

Kotoupas, A., Rigas, F., & Chalaris, M. (2007). Computer-aided process design, economic evaluation and environmental impact assessment for treatment of cheese whey wastewater. *Desalination, 213*(1-3), 238–252. doi:10.1016/j.desal.2006.03.611

Kubota, F. I., & da Rosa, L. C. (2013). Identification and conception of cleaner production opportunities with the Theory of Inventive Problem Solving. *Journal of Cleaner Production, 47*, 199–210. doi:10.1016/j.jclepro.2012.07.059

Leinweber, P., Turner, B. L., & Meissner, R. (2002). Phosphorus. In P. M. Haygarth & S. C. Jarvis (Eds.), *Agriculture, Hydrology and Water Quality* (pp. 29–55). Oxon, UK: CABI Publishing. doi:10.1079/9780851995458.0029

Lunt, J. (1998). Large-scale production, properties and commercial applications of polylactic acid polymers. *Polymer Degradation & Stability, 5*(1-3), 145–152. doi:10.1016/S0141-3910(97)00148-1

Mann, E. J. (2000). Whey products and their uses. *Dairy Industries International*, 13–14.

Marisa, H. (2004). *Degradation of whey in an anaerobic fixed bed (AnFB) reactor* (Thesis). Germany, Metcalf and Eddy. Wastewater Engineering.

Marshall, K. (2004). Therapeutic applications of whey protein. *Alternative Medicine Review, 2*, 136–156. PMID:15253675

Martin-Rilo, S., Coimbra, R. N., Martin-Villacorta, J., & Otero, M. (2015). Treatment of dairy industry wastewater by oxygen injection: Performance and outlay parameters from the full scale implementation. *Journal of Cleaner Production, 86*, 15–23. doi:10.1016/j.jclepro.2014.08.026

Mateescu, C., & Constantinescu, I. (2011). Comparative analysis of inoculum biomass for biogas potential in the anaerobic digestion. *UPB Scientific Bulletin, 73*, 99–104.

Mavrov, V., Chmiel, H., & Belieres, E. (2001). Spent process water desalination and organic removal by membranes for water reuse in the food industry. *Desalination, 138*(1-3), 65–74. doi:10.1016/S0011-9164(01)00246-6

Mawson, A. J. (1994). Bioconversions for whey utilization and waste abatement. *Bioresource Technology, 47*(3), 195–203. doi:10.1016/0960-8524(94)90180-5

Merina, U., Bernsteina, S., Bloch-Damtib, A., Yagilb, R., Van Creveld, C., Lindnera, P., & Gollop, N. (2001). A comparative study of milk serum proteins in camel (*Camelus dromedarius*) and bovine colostrum. *Livestock Production Science, 67*(3), 297–301. doi:10.1016/S0301-6226(00)00198-6

Monroy, O., Johnson, K. A., Wheatley, A. D., Hawkes, F., & Caine, M. (1994). The anaerobic filtration of dairy waste: Results of a pilot trial. *Bioresource Technology*, *50*(3), 243–251. doi:10.1016/0960-8524(94)90097-3

Najafpour, G. D., Hashemiyeh, B. A., Asadi, M., & Ghasemi, M. B. (2008). Biological treatment of dairy wastewater in an up flow anaerobic sludge-fixed film bioreactor. *American Eurasian Journal of Agricultural Environmental Sciences*, *4*, 251–257.

Overcash, M. R., & Pal, D. (1979). *Design of land treatment systems for industrial wastes-theory and practice*. Ann Arbor Science Publishers.

Polat, Z. (2009). *Integrated approach to whey utilization through natural zeolite adsorption/ desorption and fermentation* (PhD Thesis). Izmir University, Turkey.

Pouliot, Y. (2008). Membrane processes in dairy technology-from a simple idea to worldwide panacea. *International Dairy Journal*, *18*(7), 735–740. doi:10.1016/j.idairyj.2008.03.005

Prasad, P., Pagan, R., Kauter, M., & Price, N. (2004). *Eco-efficiency for the dairy processing industry. Dairy Australia.*

Qi, Y. (2008). *Effect of centrifugal dewatering on the regrowth of fecal coliforms and salmonella in anaerobically digested biosolids* (Ph.D. Dissertation). University of Delaware.

Rajeshwari, K. V., Balakrishnan, M., Kansal, A., Lata, K., & Kishore, V. V. N. (2000). State-of-the-art of anaerobic digestion technology for industrial wastewater treatment. *Renewable & Sustainable Energy Reviews*, *4*(2), 135–156. doi:10.1016/S1364-0321(99)00014-3

Sharma, D. (2014). Treatment of dairy waste water by electro coagulation using aluminum electrodes and settling, filtration studies. *International Journal of Chemical Technology Research*, *6*, 591–599.

Sienkiewicz, T., & Riedel, C. L. (1990). *Whey and whey utilization: possibilities for utilization in agriculture and foodstuffs production*. Gelsenkirchen-Buer, Germany: Verlag Th. Mann.

Siso, G. M. I. (1996). The biotechnological utilization of cheese whey: A review. *Bioresource Technology*, *57*(1), 1–11. doi:10.1016/0960-8524(96)00036-3

Smithers, G. W. (2008). Whey and whey proteins-from 'Gutter-to-Gold'. *International Dairy Journal*, *18*(7), 695–704. doi:10.1016/j.idairyj.2008.03.008

Strydom, J. P., Britz, T. J., & Mostert, J. F. (1997). Two-phase anaerobic digestion of three different effluents using a hybrid bioreactor. *Water Salination*, *23*, 151–156.

Turkmenoglu, S. (2006). *Organic acids production from cheese-whey* (Master Sci. Thesis). Middle East Technical University, Turkey.

Verheyen, V., Cruickshank, A., Wild, K., Heaven, M. W., McGee, R., Watkins, M., & Nash, D. (2009). Soluble, semi-volatile phenol and nitrogen compounds in milk processing wastewaters. *Journal of Dairy Science*, *92*(7), 3484–3493. doi:10.3168/jds.2009-2217 PMID:19528627

Vidal, G., Carvalho, A., Mendez, R., & Lema, J. M. (2000). Influence of the content in fats and proteins on the anaerobic biodegradability of dairy wastewaters. *Bioresource Technology*, *74*(3), 231–239. doi:10.1016/S0960-8524(00)00015-8

Viraraghavan, T., & Kikkeri, S. R. (1991a). Dairy wastewater treatment using anaerobic filters. *Canadian Agricultural Engineering*, *33*, 143–149.

Viraraghavan, T., & Kikkeri, S. R. (1991b). Start-up of anaerobic filters treating dairy wastewater: Effect of temperature and shock load. *Journal of Environmental Science and Health Part A.*, *26*, 287–300.

Welderufael, F., & Jauregui, P. (2010). Development of an integrated process for the production of bioactive peptides from whey by proteolytic commercial mixtures. *Separation Science and Technology*, *45*(15), 2226–2234. doi:10.1080/01496395.2010.507662

Willers, C. D., Ferraz, S. P., Carvalho, L. S., & Rodrigues, L. B. (2014). Determination of indirect water consumption and suggestions for cleaner production initiatives for the milk-producing sector in a Brazilian middle-sized dairy farming. *Journal of Cleaner Production*, *72*, 146–152. doi:10.1016/j.jclepro.2014.02.055

World Bank. (1998). *Pollution Prevention and Abatement Handbook, Dairy Industry. Draft Technical Background Document*. Washington, DC: Environment Department.

Yang, N., & Silva, E. M. (1995). Novel products and new technologies for use of a familiar carbohydrate, milk lactose. *Journal of Dairy Science*, *78*(11), 2541–2562. doi:10.3168/jds.S0022-0302(95)76884-9 PMID:8747342

Chapter 8
Impact of Trade in Class-A Environmental Products on Economic Growth and Environmental Quality:
Simultaneous Equation Analysis in Developing Countries

Sélima Ben Zineb
Independent Researcher, Tunisia

ABSTRACT

Openness to international trade enables several countries to access different markets. Similarly, the orientation of industrial productions towards ecological goods reinforces commercial activities and contributes to the improvement of environmental quality. The aim of this chapter is to estimate the indirect and the direct effects of trade in what the authors call "Class A" environmental goods on air quality for a number of developing during the period 2005-2015 (through environmental policy and income). Empirically, the study relies on the two-stage least squares (2SLS) and three-stage least squares regression analyses. For end-of-pipe (EOP) products in developing countries, neither direct nor indirect effects are identified on the reduction of pollution. For clean technologies and products, we can observe a global positive effect resulting from two positive indirect effects via environmental policy and income. In developing countries, it seems that trade in clean technologies and products (CTP) and in the OA product list generates an intensification of pollution through an increase in wealth and the adoption of a strict environmental policy (based on taxation).

DOI: 10.4018/978-1-7998-1966-0.ch008

INTRODUCTION

The World Trade Organization (WTO) emphasizes the necessity of environmental goods trade liberalization in rich countries as well as in developing countries. For all these countries, there is a way to encourage environmental protection through the use of ecological goods and the strengthening of the economic fabric. In developing countries, domestic and foreign companies benefit from a cheaper production cost based on a cheaper labor. The WTO's incentive to reduce import tariffs for environmental goods can play a significant role in encouraging business to protect the environmental quality. The reduction of tariffs encourages the imports of environmental goods and services, which in turn encourages firms to reduce pollution by benefiting from low production costs.

Reducing the costs of environmental protection products, services and technologies would encourage the governments to set up measures and programs in favour of the environmental protection.The promotion of the trade liberalization of new technologies and environmental goods favors the creation of new companies as well as new job opportunities in the domain of industrial ecology. The leading exporters of ecological goods are developed countries. They can also strengthen their economies and maintain a sustainable economic growth as far as low costs are only applicable in these countries. As stated in the work of Amwey et al. (2003), the advantages in the trade of environmental goods and services are more favorable in rich countries than in poor countries. In their work, they emphasize the importance of developed countries' membership to the World Trade Organization. All these countries benefit from gains and profits realized through trade in environmental goods. This advantage can be explained by the easy access of WTO countries to all environmental goods and services, including in poor countries.

The tariffs related to the import of environmental goods and services are profitable for the countries that do not produce this type of goods and whose industrial system is not enough developed. Even if the import of environmental goods is costly for certain countries, these latter will encourage foreign direct investments (FDI) and the implantation of new multinational corporations. The growth of Foreign Direct Investments (FDI) contributes in this case to the improvement of the production structure through technological transfer and expertise.

In the United States, all the efforts and decisions in favour of the elimination of any obstacle hindering the commercial activities of environmental goods and services are made by the business owners who are holding an important share of industrial activities. All of these business groups propose special measures for the marketing of this type of goods.

The environmental damages are largely related to the type of use of environmental goods as well as to their recycling and to their destruction (final waste disposal).

Certain countries in transition are characterized by their pollution degree. Environmental degradation can be explained by the consumption process of goods and their elimination in the environment.

The question now being asked is: what needs to be liberalized in developing countries in order to preserve the quality of the environment?

This paper will be presented as follows: the first part includes a short review of the literature on environmental goods. It will consist of a presentation and a classification of environmental goods. The second part will deal with the presentation of our theoretical model which is based on a specification of three simultaneous equations. The variables of this study will be defined, followed by an estimation approach. Finally, the results of the estimations will be the subject of the third part in which the results will be analyzed and discussed. The paper will conclude with some recommendations.

REVIEW OF THE LITERATURE

Classification of Environmental Goods

A classification that categorizes environmental goods and all the existing subcategories within each group of goods has been used for all the goods mentioned in this study. Although UNCTAD provides a classification of the goods that are traded in terms of export and import volumes, we decided to rely on a more detailed classification known as "Harmonized Commodity Description and Coding System s" (HS). In this work, the retrieval of environmental goods and Environmentally Preferable Products (EPP) was based on the classification of the Harmonized System-96 (HS-96 6- digit HS code), (UNCTAD, U., 2003).

As already mentioned, the classification of product subcategories is carried out by a digit code which renders the description of all goods clearer. The order of codification of goods is composed of only six numbers, as used by the World Trade Organization. The more the number of the harmonized code exceeds the number six, the more the chances to get a thorough representation of the products will be high. Within the six-digit code sub-categories, there are certain products whose consumption does not have environmental protection as main purpose. For a good reason, some products such as pipes can be used as environmental protectors (commonly used in wastewater treatment activities) or as normal products whose purpose is not related to the environment (UNCTAD, U., 2003).

To define and describe the set of environmentally preferable products, the classification list of environmental goods established by UNCTAD has been used. Indeed, some goods and products do not have particular characteristics but their production was based on a set of techniques and practical methods that take into account environmental standards. To extract these goods, it is important to use certified products, such as water-based paint products that are deprived of any type of solvent (...).Several environmentally preferable products have identical functions. The WTO does not identify them and does not classify them according to a particular list. It rather takes into consideration only one criterion based on "what the good represents" for the environment and its consumption.

Based on a criterion related to the final consumption of a good, a list was created in 1997 by the OECD to clearly define all these products. In its work, the OECD relies on a set of goods and products that possess a strong capacity to measure environmental damage. Similarly, the OECD lays emphasis on the capacity of a good to reduce environmental impacts and to ensure the best management of existing resources. In this sense, environmental goods are identified at the same level as the products whose classification is indexed with a six-digit code. The harmonized system encompasses all kinds of marketed products and does not separate the category of environmental goods and services from other products. As a result, following a rather important codification, a new list of environmental goods was developed by the Asia-Pacific Economic Cooperation (APEC) to discuss the custom duties of the products covering the period 1998 - 2000. In this list, the products that appear in the subcategories of goods are classified according to 8 or 10-digit codes. Although there are two lists (OECD and APEC) that classify environmental goods, we are not yet able to confirm their final and exact classification as the lists are still not complete. The technical progress and the continuous development of new technologies are leading to the emergence of new advanced products that can be classified as environmental goods. This progress requires a constant update of the lists in order to take into account new products and developments in different fields. Despite the efforts of several global organizations to provide a clear and accurate list of environmental goods, the works of the WTO are mainly based on environmental goods defined by the "Class A" list belonging to the OECD and the APEC . As an example, we can cite the attempt of the UNCTAD which presented a new list named "EPP-core" whose products are similar to those categorized in the list of goods preferable for the environment (Class B). All the goods in the "EPP-core" list consist of products designed for industrial use and individual consumption. It is the WTO, after few debates on the trade of environmental goods, which designated the products of this list as preferablė goods for the environment according to their use and their specific consumption.

It should be noted that our subsequent estimate will only focus on the environmental goods of the so-called Class A.

Previous Studies

No research has been conducted yet on the causal relationship between trade liberalization in environmental goods and services and their effects on countries' economic growth and environmental quality. However, in her study, Zugravu-Soilita (2009) developed a framework of analysis that will serve as a model for the present study. The results of my study will then be compared to the results of her analysis.

On the subject of trade liberalization, some studies addressed how trade in environmental goods and services influences the implementation of policies protecting the environment (Feess & Muehlheusser (1999, 2002), Copeland & Taylor (2005), Greaker & Rosendahl (2006), Canton (2007)). However, the effects of trade in environmental goods and services on countries' economic growth and environmental quality have been disregarded.

Feess & Muehlheusser (1999, 2002), Copeland & Taylor (2005), Greaker & Rosendahl (2006), Canton (2007) focused on the influence of trade in environmental goods and services on the implementation of environmental protection policy.

In their analysis, Bora & Teh (2004) justified the importance of the liberalization of environmental goods. In fact, trade in this type of goods increases the energy efficiency of importing and exporting countries and allows them to reduce pollutant emissions.

Such a result consolidates the idea presented by the WTO which attests to the importance of trade in this type of goods. The reduction of pollution can be explained by the right use of all environmental goods and the existence of other constitutive elements with indirect effects influencing the relationship between air pollution and the use of products reducing environmental pollution. Among the indirect factors influencing the reduction of polluting emissions, we can mention the importance of the environmental policy to be implemented. Since all economic activity creates pollution, the strengthening of environmental policy could reduce emissions. It is therefore considered as a fundamental element as it indirectly plays a significant role in the reduction of polluting emissions subsequent to the use and marketing of environmental goods.

The role of policy-makers consists in setting the right environmental policy while taking into consideration countries' entry into international markets and the need to reduce environmental damage. Countries' development and economic growth also have considerable mitigating effects on pollution. Developed countries are famous for their advanced techniques and technologies, which are available to increase social welfare and develop the industrial sector. For this reason, it goes without saying that technical progress plays an important role in offsetting the scale effect.

Any pollution generated as a result of the increase in the scale of production is reduced through the improvement of production processes and techniques ensuring a better quality of the environment. In addition, economic development contributes to reducing environmental damage through stable and effective environmental policies. In this context, the effectiveness of the environmental policy is a key element for the technical effect to lead to environmental improvement.

Technical effect is realized through the increase of income level through the accumulation of wealth. Raising the level of income therefore leads individuals to make some sacrifices in order to protect their environment and consume goods and services that do not cause environmental damage. In view of the fact that development is based on the level of income, any tax exemption in countries whose business activity is based on imports of environmental goods and services causes a decline in the level of income. For this reason, the environmental component cannot be made a priority.

We will try to focus on the effects that trade in environmental goods and services can have on the quality of the environment while taking into account the different elements influencing the improvement of the quality of the environment.

The commercial movements of environmental goods can be determined and influenced by various factors such as the effectiveness of the environmental policy and its sustainability, the obstacles in the economic environment, the evolution of people's needs and their demand for biological goods called "green" products and the new direction of companies towards new ecological technologies.

In the following sections, we will make an estimate of the direct and indirect effects of the trade liberalization of environmental goods on the quality of the environment according to a model based on three specific simultaneous equations. Such an estimate will take into consideration the previously mentioned elements that have an indirect influence on the trade of environmental goods. The indicator of environmental degradation chosen in this study is analyzed and explained in the first equation. We then focus on the theoretical framework developed in the work of Grossman (1995) as well as Taylor & Copeland (2001).

In the second equation, we explain the environmental policy that is implemented and its effect on the sustainability and stability of the environment by focusing on some theoretical frameworks emphasizing the importance of such an indicator (Petrović et al. (2003)). Finally, for the third equation which aims to explain the variable of wealth expressed in terms of the per capita income variable, we rely on some theoretical analyses within the scope of the endogenous growth theory [[1]].

EMPIRICAL VALIDATION

Estimation Model and Methodology

For a country "p" during a period (t), pollution is determined below by several factors presented as follows:

$$Pe_{pt} = f\left(VA_{pt}, va_{pt}, i_{(gross)pt}, PE_{pt}, Trade_{pt}\right) \tag{1}$$

The added value achieved by each specific sector (va_{pt}) depends in particular on the factors of production (capital endowments), labor and capital (capital stock) on which the composition effect depends. The labor factor is related to the entire working population, also called labor force. The added value (VA_{pt}) is relative to the overall added value for the whole economy. The variable (PE) expresses the demand for clean products to ensure the clean-up process. The variable (Trade) relates to a country's openness to international trade, and is expressed in terms of open rate.

The level of gross intensity of polluting emissions $(i_{(gross)p})$ is expressed in terms of the production tools used (Clean technologies and products (CTP)). It also depends on the Environmental Policy Stringency (the severity of environmental policy) implemented and the income level reached in an economy. Below is the specification of the gross pollutant intensity for a country "p" during a period of time (t), illustrated as follows:

$$i_{(gross)pt} = f\left(CTP_{pt}, P_{env(pt)}, Rev_{pt}\right) \tag{2}$$

In this specification, the factor (CTP) refers to trade intensity in ecological products and most advanced "clean" technologies. We also emphasize the effectiveness and importance of environmental protection policy. (P_{env}). (Rev_{pt}) refers to the per capita income level achieved in the economy which determines the capacity, the awareness of individuals and their commitment to protect the environment.

The technical effect is justified by considerable efforts made to reduce pollution and by the demand for "end-of-pipe" technologies (EOP) to ensure the clean-up process, as specified below:

$$PE_{pt} = f\left(EOP_{pt}, P_{env(pt)}, Rev_{pt}\right) \tag{3}$$

The demand for depollution is expressed in terms of trade intensity in end-of-pipe technologies.

Based on theoretical assumptions, the coefficient of the parameters expressing the composition effect and the scale effect are expected to have a positive effect on pollution. However, given their importance in improving the environment, we expect the coefficients of the variables that express the technical effect to be negatively correlated with the level of pollution.

The Environmental Equation

According to Greaker & Rosendahl (2006), the application of a rigid and strict environmental policy is not beneficial for industries exporting environmentally preferable goods and clean technologies. Similarly, for companies whose production activities are highly polluting, strict policies are likely to hinder their activities. In this sense, several companies are encouraged to relocate their activities to countries where environmental regulations are not rigid. The effect of trade intensity in "Clean Technologies and Products" (CTP) as well as "end-of-pipe" (EOP) products on pollution is still ambiguous.

The specification of the environmental policy *(P_{evn})* function and its key determinants are illustrated as follows:

$$P_{env(pt)} = f\left(Freed\_index_{pt}, Corrp\_Cntr_{pt}, POLITstab_{pt}, EOP_{pt}, CTP_{pt}, Trade_{pt}\right) \tag{4}$$

The term (Freed_index) is an indicator of democracy; (Corrup_Cntr) refers to corruption, and (POLITstab) expresses the political stability of a country. The other terms have been previously defined as (EOP), which is related to "end-of-pipe" products (EOP); (CTP) which expresses "clean" products and clean production technologies. Finally, (Trade) is related to trade openness.

The Income Equation

In order to define and clarify the various factors and determinants of the income function, we rely on an endogenous growth analysis found in the literature. Production depends on the capital stock and labor as well as countries' level of development. According to Rodrik et al. (2004), the studies of Acemoglu & Robinson (2001) and Sachs (2003) focused on the "institutional" and "geographic" [2] factors and the importance of "international trade" in determining the factors of income level.

In our analysis, we only focus on the institutional factor and the role of trade liberalization in the development of nations. In fact, the establishment of better institutions and the effectiveness of a country's political structure lead to its social and economic development (Subramanian and Trebbi, 2004). All these points have been developed thoroughly following a rich literature which justifies the role of all these factors in the economic development [3]. International trade is seen as a factor of economic growth because of its positive and significant contribution to improving income level (Frankel & Romer (1999)).

In our case study, the different determinants for a country (p) on which income *(Rev)* depends are illustrated as follows:

$$Rev_{pt} = f\left(K_{pt}, L_{pt}, Freed\_index_{pt}, EOP_{pt}, CTP_{pt}, Trade_{pt}\right) \tag{5}$$

The terms K and L refer to the factors of production: the stock of available capital and labor. The quality of the institutions in this case is expressed in terms of the nature of the democratic regime that has been adopted. It is represented by the variable of civil liberties and political rights.

Our empirical validation will valorize the different effects of trade intensity in environmental goods on the quality of the environment. It is developed through a system of three simultaneous equations. Trade in environmental goods has a direct effect on the level of pollution and will be illustrated in the first equation. However, the second equation highlights the indirect effect that trade in environmental goods can have through environmental regulation and policy implementation. Finally, the third equation serves to test the indirect effect of trade in environmental goods through income.

The terms K and L refer to the factors of production: the stock of available capital and labor. The quality of the institutions in this case is expressed in terms of the nature of the democratic regime that has been adopted. It is represented by the variable of civil liberties and political rights.

Our empirical validation will valorize the different effects of trade intensity in environmental goods on the quality of the environment. It is developed through a system of three simultaneous equations. Trade in environmental goods has a direct effect on the level of pollution and will be illustrated in the first equation. However, the second equation highlights the indirect effect that trade in environmental goods can have via environmental regulation and policy implementation. Finally, the third equation serves to test the indirect effect of trade in environmental goods via income.

Since the sum of the added values of the different production sectors represents the scale of the economy expressed as a function of the Gross Domestic Product (GDP), our final structural model is presented as follows (*Pe* is relative to pollution level):

$$\begin{cases} Pe_{pt} = f\left(GDP_{pt}, Rev_{pt}, P_{env(pt)}, EPP-core_{pt}, CT_{pt}, Trade_{pt}\right) \\ P_{env(pt)} = f\left(Rev_{pt}, Freed\_index_{pt}, Corrp\_Cntr_{pt}, POLITstab_{pt}, EPP-core_{pt}, CT_{pt}, Trade_{pt}\right) \\ Rev_{pt} = f\left(K_{pt}, L_{pt}, Freed\_index_{pt}, EPP-core_{pt}, CT_{pt}, Trade_{pt}\right) \end{cases} \quad (6)$$

In this model, the variables of pollution, environmental policy and realized income are dependent variables.

Variables of the Model

The sample of developing countries in this study includes 21 low-income countries and 37 middle-income countries. Our estimate will cover the period 2005 to 2015 insofar as the indicator of environmental sustainability, which is expressed according to the effectiveness of the adopted environmental policy, is only available during this period. In order to estimate our model, we prepared a database that illustrates the different factors influencing the quality of the environment. In this sense, we presented a theoretical model based on three key dependent variables: pollution, Environmental Policy Stringency and income level. The definition of the variables and the sources of data extraction are presented in **table 6** of the appendix.

Estimation Technique

If we need to identify the direct and indirect effects of trade liberalization on the quality of the environment, we should opt for estimation methods that can clearly identify these effects. If we use the OLS method (Ordinary Least Square), the estimated results would only show a partial effect. In order to choose the adequate estimation method and valorize the different trade effects on the quality of the environment, we do estimations through Two-Stage Least Squares (2SLS) and Three Stage Least Squares regression analysis. For a more adequate method of estimation, we rely on the Hausman test results which are illustrated in the estimation tables.

RESULTS ESTIMATES

Analysis of the Equation Related to Environmental Quality

By analyzing the results of the first equation representing the Quality of the Environment (**Table 1**), we observe that most of the results verify the predictions of the theoretical model.

For the scale of economic activity represented by the Gross Domestic Product variable (GDP), we observe two positive and significant coefficients in model (1) and Model (2) at the 5% level and at the 10% level respectively. As a result, an increasing volume of large-scale production contributes to the massive increase in CO_2 emissions in developing countries.

In the two models, the coefficients of the variable (SPE: Strenght of Environmental Policy) representing the environmental regulatory policy have positive and significant signs (unlike the predictions of the theoretical model) at the 5% level and at the 10% level respectively. The environmental policies implemented in developing countries do not contribute to the reduction of polluting emissions. It is the trade policy of product diversification that can be effective in reducing pollution and for this reason, diversified exports must be encouraged. This issue can be explained by the low level of "severity" related to regulations and standards in favor of the protection of the environment. Developing countries suffer from a lack of means and techniques to implement the most rigorous and environmentally beneficial policies. If developing countries choose to set rigid standards to reduce carbon emissions through the adoption of a strict policy, this will lead to the loss of competitiveness of local industries as the majority of industrial and manufacturing poor countries are highly polluting. In this context, two possible alternatives are to be adopted. Some companies are relocating to countries with lower restrictions, whereas some major industries choose to increase their business by expanding their investments in countries that want to achieve growth rate. This will subsequently contribute to the creation of wealth in the country and weaken environmental policies. For countries that want to achieve high growth rate, some major industries choose to increase their business by expanding their investments; which will contribute to the creation of wealth in the country and weaken environmental policies. This phenomenon is known as the scale effect, leading to an intensification of polluting emissions.

Moreover, the results obtained for gross national income are contrary to theoretical predictions. We note that an increase in national income leads to an intensification of pollution. This is illustrated by positive and significant coefficients of the variable (GNPp) at the 1% level.

Trade liberalization has a direct positive effect on the level of carbon emissions generated. The results of Model (2) illustrate a positive and significant coefficient at the 1% level. However, trade in services seems favorable for the environment and decreases pollution. As a result, we observe two negative and significant coefficients at the 5% level. Results showed that air pollution decreases with an increasing trade volume in ICT (New Information and Communication Technologies), as we obtained negative and significant sign coefficients at the 1% level in both models.

In the case of environmental goods in Model (1), we observe a positive and non-significant coefficient of trade intensity (TradeINT_OA / A) for the products of the OA list. As a result, for OA products there is no direct effect of trade intensity on the level of pollution in developing countries.

This can be explained by low volumes of imports in environmental goods in low and middle-income countries because of high import tariffs. Similarly, for the two categories of goods in the OA list (Model (2)), the estimated coefficients for the following variables: "end-of-pipe products" and "clean technologies and products" are non-significant and do not reflect any direct effect on the level of pollution. We obtained a negative but not significant coefficient for the CTP trade intensity variable. Similarly, any increase in the per capita income level leads to an increase in CO2 emissions.

In Model (1), we observe two coefficients of concentration indices for imported products with positive and significant signs at the 1% level. The Concentration of imports in certain products increases air pollution. This result can be explained by the trade of developing countries which is mainly based on the import of polluting products from developed countries. However, the coefficients of concentration indices for exported products are negative and statistically significant, which explains that imports made by developing countries are mainly based on goods whose production and destruction are polluting.

This further shows that the concentration on a homogeneous group of imported products is not favorable for the environment of developing countries. However, the diversification of exported products has a negative effect on the environment, which was justified by the results of Model (2) with a negative and significant coefficient of the variable (IDiv_exp) at the 10% level. The diversification of products thereby decreases pollution over time. In order to increase production and expand economic activity, countries cannot focus solely on high-technology products and ecological activities. In this sense, increasing economic activity requires a concentration of polluting activities, which means that the concentration of products generates CO_2 emissions. It is therefore important for developing countries to opt for diversified trading activities to protect the environment and reduce imports based on polluting goods and equipments.

Table 1. Estimation of the direct effect of the trade intensity in environmental goods on CO_2 emissions by DMC and TMC techniques in Panel

Equation (1): ***Estimation of the direct effect on environmental quality***	**Developing countries (middle-income and low income countries)**	
	Model (1)	**Model (2)**
	TMC	**TMC**
	ln(ECO_2p)	***ln(ECO_2p)***
ln(GDP)	0.40959** (2.06)	0.40417* (1.96)
ln(K/L)	-0.21210** (-2.30)	-0.20934** (-2.13)
ln(GNP_p)	1.16247*** (9.11)	1.16829*** (8.80)
ln(ISEP)	1.18703** (2.00)	0.97064* (1.72)
ln(TradeINT_OA/A)	0.00552 (0.09)	0.03327 (0.49)
ln(TradeINT_CTP/OA)		-0.05634 (-0.74)
ln(Trade_ICT)	-0.34104*** (-4.92)	-0.34287*** (-5.03)
ln(Trade_SERV)	-0.36643** (-2.48)	-0.34588** (-2.37)
ln(Trade)	0.88031 (6.25)	0.84385*** (6.12)
IC_imp	2.25340*** (2.86)	2.30225*** (2.94)
IC_exp	-0.77496*** (-2.62)	-0.76900** (-2.46)
IDiv_imp	-1.49358 (-1.63)	-1.51169 (-1.62)
IDiv_exp	-1.29610* (-1.81)	-1.33482* (-1.90)
Constant	-13.34179*** (-10.15)	-12.98644*** (-9.9)
Hausman test: Prob>Chi2	0.1314	0.0676
Observations number	246	246

Notes: (***): Significance at the 1% level; (**): significance at the 5% level and (*): significance at the 10% level. Values in parentheses are relative to « t » of student. **DMC**: Two-stage Least Squares Technique. **TMC**: Three-stage Least Squares Technique.

According to the estimation of our first equation, we can conclude that ICT trade is good for the environment in developing countries. Similarly, trade in services is favorable and reduces pollution. Trade in environmental goods (OA list) has no direct effect on the level of carbon emissions. We can not confirm a clear result for the trade intensity in EOP products and green technologies since they do not have a direct impact on pollution.

All these identified results are not sufficient to reach a final conclusion because we must take into consideration indirect effects in our estimates.

Analysis of the Equation Related to the Environmental Policy Stringency

After examining the results identified in the estimation of the second equation related to environmental regulation in Model (1) and (2) in **Table 2**, we observe positive and significant signs for the coefficients of the national income variable at the 1% level. As a result, increasing the level of national income has a positive effect on the Environmental Policy Stringency in the case of CO_2 emissions in developing countries.

We also illustrate results that verify our theoretical model for the case of the institutional environment. We identify positive and significant signs for the coefficients of the political stability at the 5% and 10% level respectively (case of the two models). This means that a stable political environment generates a better environmental policy. Similarly, for developing countries, political stability encourages foreign direct investment and trade, which enables them to benefit from the import of environmentally friendly tools and products. For the control of corruption, we identify results that go in line with theoretical expectations, as the control of corruption increases the Environmental Policy Stringency. Civil liberties contribute positively to the effectiveness of the environmental policy. Individuals will be encouraged to protect the environment in the presence of a democratic political regime through their participation in social activities that strengthen the efforts of pressure groups in the protection of the environment, (You et al. (2015)).

For the international trade variable represented by the trade opening variable, we identify in Models (1) and (2) two negative and significant sign coefficients at the 1% level. In this context, we can confirm that the impact of global trade on environmental policy results from the "race to the bottom" phenomenon. The intensification of trade liberalization reduces the Environmental Policy Stringency. In fact, the improvement of economic activity and the increase of investment capital to improve the commercial fabric are carried out at the expense of the policies and regulations applied to protect the environment, which leads to their depreciation.

Table 2. Estimation of the indirect effect of the trade intensity in environmental goods on CO_2 emissions via environmental policy by DMC and TMC techniques in Panel

Equation of the environmental policy (2): Estimation of the indirect effect via environmental policy	**Developing countries (*middle-income and low income countries*)**	
	Model (1)	**Model (2)**
	TMC	**TMC**
	ln(ECO₂p)	***ln(ECO₂p)***
ln(ISEP)		
ln(GNP_p)	0.055215*** (2.77)	0.05767*** (2.87)
POLITstab	0.03367** (2.02)	0.03243* (1.93)
Freed_index	0.03112*** (3.35)	0.03523*** (3.68)
Corrp_Cntr	0.11647*** (4.41)	0.10965*** (4.10)
ln(TradeINT_OA/A)	-0.00428 (-0.23)	
ln(TradeINT_ETP/OA)		-0.02942 (-1.50)
ln(TradeINT_CTP/OA)		0.02041 (0.90)
ln(Trade_ICT)	0.04101* (1.91)	0.04062* (1.90)
ln(Trade_SERV)	0.14414*** (5.25)	0.14507*** (5.28)
ln(Trade)	-0.21113*** (-6.40)	-0.21267*** (-6.46)
IC_imp	0.08190 (0.35)	0.086317 (0.37)
IC_exp	0.28338*** (3.85)	0.30732*** (4.07)
IDiv_imp	0.04199 (0.16)	-0.05189 (-0.19)
IDiv_exp	-0.38859* (-1.93)	-0.42357** (-2.10)
Constant	1.06124*** (3.98)	1.02665*** (3.84)
Hausman test: Prob>Chi2	0.1314	0.0676
Observations number	246	246

Notes: (***): Significance at the 1% level; (**): significance at the 5% level and (*): significance at the 10% level. Values in parentheses are relative to « t » of student. **DMC**: Two-stage Least Squares Technique. **TMC**: Three-stage Least Squares Technique.

The intensity of trade in environmental goods, and more precisely the intensity of trade in goods appearing in the OA list has no effect on the Environmental Policy Stringency. Similarly, in the two categories of goods appearing in the OA list (Model 2), the estimated coefficients of the variables "end-of-pipe products" (EOP) and clean technologies and products are insignificant and present no direct effect on the Environmental Policy Stringency. Trade in ICTs and services reinforces the Environmental Policy Stringency. The technologies that are developed and the services favor the success of environmental initiatives since the activities operating in these markets act in conformity with the standards fixed by the state.

Analysis of the Income Equation

By comparing our estimated results in the third income equation (**table 3**) with what is developed in the endogenous growth theory, we could identify a robust coefficient of the variable which is representative of the capital endowments and consistent with theoretical assumptions. The variable (K) defining relative capital endowments has a positive and significant effect on the level of wealth achieved. [4]

The intensity of trade in the environmental goods of the full OA list has a positive and significant effect on the level of income. This is justified by a coefficient of the variable (TradeINT_OA / A) of positive and significant sign at the 5% level. The intensity of EOP trade has no effect on the level of wealth. In addition, the intensity of trade in clean technologies and products has a positive and significant effect on income. We observe a coefficient of the variable (TradeINT_CTP / OA) of positive and significant sign at the 10% level. Civil liberties and public rights have a positive and significant impact on the level of income achieved (positive and significant coefficients at the 1% level).

As demonstrated in the theoretical model and in the work of Frankel & Romer (1999) in the case of international trade, the results illustrated show that any increase in trade openness generates increased income in developing countries (and vice versa). This is a robust result as Frankel & Romer (1999) found that trade liberalization increases the level of wealth.

Similarly, we have included diversification and concentration variables in exports and imports in the income equation. In fact, we observe that the concentration of exports and imports has no effect on income. Concerning diversification, we obtain coefficients of import diversification indices of negative and significant signs at the 5% and 10% levels. As a result, the diversification of imports decreases income. Imported products will compete with local products which will lower their prices due to competition.

In what follows, we will summarize our results by determining the overall effect of the intensity of trade in environmental goods in developing countries (through the

Table 3. Estimation of the indirect effect of the trade intensity in environmental goods on CO_2 emissions via income by DMC and TMC techniques in Panel

Équation of income (3): Estimation of the indirect effect via the income level	***Developing countries (middle-income and low income countries)***	
	Model (1)	**Model (2)**
	TMC	**TMC**
	ln(ECO₂p)	***ln(ECO₂p)***
ln(GNP$_p$)		
ln(K)	0.76043*** (-1.05)	0.76469*** (23.52)
ln(L)	-0.65241*** (-20.82)	-0.65286*** (-20.84)
Feed_index	0.04734*** (2.76)	0.04932*** (2.80)
ln(TradeINT_OA/A)	0.08354** (2.22)	
ln(TradeINT_EOP/OA)		0.01080 (0.27)
ln(TradeINT_CTP/OA)		0.09064* (1.89)
ln(Trade_ICT)	-0.04363 (-1.06)	-0.04279 (-1.04)
ln(Trade_SERV)	0.08354** (2.22)	0.08992 (1.38)
ln(Trade)	0.16866*** (2.66)	0.16991*** (2.68)
IC_imp	-0.48917 (-1.05)	-0.54599 (-1.17)
IC_exp	0.06619 (0.45)	0.09254 (0.61)
IDiv_imp	-1.13635** (-2.10)	-1.04790* (-1.87)
IDiv_exp	0.67181 (1.56)	0.67275 (1.56)
Constant	0.75272 (0.95)	0.64915 (0.79)
Hausman test: Prob>Chi2	0.1314	0.0676
Observations number	246	246

Notes: (***): Significance at the 1% level; (**): significance at the 5% level and (*): significance at the 10% level. Values in parentheses are relative to « t » of student. **DMC**: Two-stage Least Squares Technique. **TMC**: Three-stage Least Squares Technique.

direct effect and the indirect effects via the regulation of the environment and the amount of realized income). We will determine whether the indirect effects tend to decrease or offset the direct effect of trade intensity in environmental goods on the level of pollution. We will then show in Table 6 of the appendix the overall effect obtained after calculating it [5].

Similarly, for all variables, the calculation method [6] of the total effect is as follows:

$Coeff(TradeINT\_EOP/OA)_{Eq(1)}$ → ***Direct effect(1)***

$[Coeff(TradeINT\_EOP/OA)_{Eq(2)} * Coeff(ISEP)_{Eq(1)} + Coeff(TradeINT\_EOP/OA)_{Eq(3)} * Coeff(GNPp)_{Eq(2)}]$ → ***Indirect effect via the policy of the environment (2)***

$[Coeff(TradeINT\_EOP/OA)_{Eq(3)} * Coeff(GNPp)_{Eq(1)}]$ → ***Indirect effect via income (3)***

→ *(1) + (2) + (3) = Total effect*

Analysis of the Overall Effect of Trade Intensity in the Environmental Goods of the OA List on Pollution

By analyzing the results shown in Table 6 of the appendix, we observe a positive overall effect resulting from two positive indirect effects via income and the Environmental Policy Stringency. No direct effect has been identified in the trade of OA products. In this case, the indirect effects related to income and environmental policy are not considered as technical effects since they create a positive total effect. We must therefore opt for a more flexible environmental policy. Concerning income, unlike the results of other studies which showed that income is an essential factor to the reduction of pollution, our study shows that an increase in wealth increases pollutant emissions. Thus, a high intensity of trade in OA products increases air pollution through increased income. After having shown the positive indirect effect of income and environmental policy we can conclude that trade in OA goods has a positive overall effect.

Similarly, for all variables, the calculation method [7] of the total effect is as follows:

$Coeff(TradeINT\_EOP/OA)_{Eq(1)}$ → *Direct effect(1)*

$[Coeff(TradeINT\_EOP/OA)_{Eq(2)} * Coeff(ISEP)_{Eq(1)} + Coeff(TradeINT\_EOP/OA)_{Eq(3)} * Coeff(GNPp)_{Eq(2)}]$ → *Indirect effect via the policy of the environment (2)*

$[Coeff(TradeINT\_EOP/OA)_{Eq(3)} * Coeff(GNPp)_{Eq(1)}]$ → *Indirect effect via income (3)*

→ *(1) + (2) + (3) = Total effect*

Table 4. Global Effect of Trade Intensity in Environmental Goods

Trade intensity	Air pollution: $PECO_2$			Global effect = direct effect + indirect effects
	Direct effect	Indirect effect via:		
		(ISEP)	(GNP_p)	
(Goods of the OA list) **(TradeINT_OA/A)** Model (1)	-	**(0.055215x0.08354) =0.00461 (+)**	**(0.08354x1.16247) =0.10291 (+)**	**(0.10752) (+)**
"End of Pipe" product TradeINT_EOP/OA Model (2)	-	-	-	-
Clean technologie and product ***TradeINT_CTP/OA*** **Model (2)**	-	**(0.09064 x 0.05769) = 0.00522 (+) ISEP→GNPp →CPT**	**(0.09064x1.16829) =0.10588(+)**	**(0.11110) (+)**

Notes: The effects are calculated through the coefficients identified in the estimates of the three simultaneous equations based only on the significant coefficients of the trade intensity variables related to the following Class A environmental goods:: *TradeINT_OA/A, TradeINT_EOP/OA, TradeINT_CTP/OA*, the variables of ISEP and income.

Analysis of the Overall Effect of Trade Intensity in "End-of-Pipe" Products (EOP) on Pollution

We did not identify any effects on CO_2 emissions as income and the Environmental Policy Stringency did not have any significant indirect effect on pollution.

- By analyzing the results shown in Table 6, we observe a positive overall effect resulting from a positive indirect effect via income and another positive indirect effect through environmental policy (the second effect includes the effect of the Environmental Policy Stringency on income). No direct effect of trade in "Clean Technologies and Products" (CTP) is identified. In this case, the effect of income and the effect of environmental policy are not considered as technical effects. They led to an intensification of air pollution through trade in "Clean Technologies and Products" (CTP). As a result, increased wealth leads to increased CO_2 emissions. Similarly, a severe policy causes a harmful effect on the environment of developing countries.

It is therefore necessary to choose the optimal environmental policy (other than taxation) that leads to the reduction of air pollution. For any increase in revenue, it is necessary to focus on R & D (research and development) activities and to install high-tech tools that can produce with the lowest level of pollution possible. On the one hand, the lack of a clear effect of trade in environmental goods in developing countries can mainly be explained by the choice of the study period which started in 2005. On the other hand, it can be explained by the market failure of these products during the last period. Some developing countries and countries in transition dominated the environmental goods market in 2000 and 2001 with growth rates of 7% and 8%, respectively.

Although markets for environmental goods have been dominated by a few developing countries, many of these countries have not yet had access to more developed product and technology markets. In addition, high import tariffs are an essential factor in determining the intensity of trade in environmental goods. In fact, Bora & The (2004) showed in a graphic illustration that import tariffs for environmental goods in developing countries are very high by comparing them to that of developed countries whose fixed taxes are considered exorbitant. It is necessary for developing countries to encourage trade in environmentally friendly goods by removing barriers to the import of these goods. It is therefore necessary to implement a more flexible environmental policy and encourage investment in technological innovations and the installation of more efficient production equipment.

As long as the commercial activity is mainly based on imports and exports, and as long as the results illustrated in this estimate do not give a clear idea of the origin of certain direct effects, it is necessary to estimate the effects generated by import and export flows for each type of environmental good.

CONCLUSION

In the case of developing countries, we found a positive overall effect in the OA list resulting from a positive indirect effect through income and environmental policy. As a consequence, the stringency of environmental policy generates air pollution through trade in the goods of the OA list. The flexibility of environmental regulations can fix the situation while acting on other factors to reduce pollution.

Moreover, the additional income generated by countries has to be reinvested in technological innovations. For EOP products, we did not identify any direct or indirect effect on pollution. In the case of CTP products, we observed a general positive effect resulting from two indirect positive effects via environmental policy and income. In developing countries, it seems that trade in CTP and in products of the OA list generates an intensification of pollution through an increasing wealth and the adoption of a strict environmental policy (based on taxation). In fact, the dominance in the market of environmental goods in developing countries did not last.

In addition, these results can be explained by the unavailability of diversified clean products in certain countries. Similarly, high tariffs on imports play a major role in the impact of trade in "Class A" products on pollution.

This high taxation hindered trade in diversified environmental products. Developing countries are therefore prompted to encourage trade in environmental products by eliminating import barriers in order to expand the circle of products and benefit from a variety of environmental goods. The indirect effect of income in developing countries does not seem beneficial to minimize environmental degradation. Any increase in wealth leads to an increase in carbon emissions. It is therefore necessary for developing countries to increase technical progress and to opt for technological innovation by focusing on research and development to offset the harmful effects of increased wealth. In addition, choosing a more flexible environmental policy can increase the benefits of reducing pollution.

New entrepreneurs need to take into consideration the quality of the environment in the creation of new products. The notion of "green" products is expressed in the context of the definition of ecological products and their importance in improving the environment. A new line of research should be further explored with a new question opening up perspectives on the role of entrepreneurship, and specifically 'green' entrepreneurship in improving the environment. Now the question remains: *how can new entrepreneurs create new products which have the ability to reduce environmental degradation?*

REFERENCES

Acemoglu, D., & Robinson, J. A. (2001). A theory of political transitions. *The American Economic Review*, *91*(4), 938–963. doi:10.1257/aer.91.4.938

Bora, B., & Teh, R. (2004, October). Tariffs and trade in environmental goods. In *Workshop on Environmental Goods* (*Vol. 11*). Academic Press.

Canton, J. (2007). *Environmental taxation and international eco-industries*. Academic Press.

Copeland, B. R., & Taylor, M. S. (2005). Free trade and global warming: A trade theory view of the Kyoto protocol. *Journal of Environmental Economics and Management*, *49*(2), 205–234. doi:10.1016/j.jeem.2004.04.006

Feess, E., & Muehlheusser, G. (1999). Strategic Environmental Policy, International Trade and the Learning Curve: The Significance of the Environmental Industry. *Jahrbuch für Wirtschaftswissenschaften. Revista de Economia (Curitiba)*, 178–194.

Feess, E., & Muehlheusser, G. (2002). Strategic environmental policy, clean technologies and the learning curve. *Environmental and Resource Economics*, *23*(2), 149–166. doi:10.1023/A:1021249404533

Frankel, J. A., & Romer, D. H. (1999). Does trade cause growth? *The American Economic Review*, *89*(3), 379–399. doi:10.1257/aer.89.3.379

Greaker, M., & Rosendahl, K. E. (2006). *Strategic climate policy in small, open economies*. Academic Press.

Grossman, G. M. (1995). Pollution and growth: what do we know. *The Economics of Sustainable Development, 19*, 41.

Hamwey, R., Hoffman, U., Vikhlyaev, A., & Vossenar, R. (2003). *Liberalization of international trade in environmental goods and services*. Bangkok: UNCTAD.

Mankiw, N. G., Romer, D., & Weil, D. N. (1992). A contribution to the empirics of economic growth. *The Quarterly Journal of Economics*, *107*(2), 407–437. doi:10.2307/2118477

Petrović, M., Gonzalez, S., & Barceló, D. (2003). Analysis and removal of emerging contaminants in wastewater and drinking water. *TrAC Trends in Analytical Chemistry*, *22*(10), 685–696. doi:10.1016/S0165-9936(03)01105-1

Rodrik, D., Subramanian, A., & Trebbi, F. (2004). Institutions rule: The primacy of institutions over geography and integration in economic development. *Journal of Economic Growth*, *9*(2), 131–165. doi:10.1023/B:JOEG.0000031425.72248.85

Sachs, J. D. (2003). *Institutions don't rule: direct effects of geography on per capita income (No. w9490)*. National Bureau of Economic Research. doi:10.3386/w9490

Taylor, M. S., & Copeland, B. R. (2001). Trade and transboundary pollution. In *The Economics of International Trade and the Environment* (pp. 125–148). CRC Press. doi:10.1201/9781420032628.ch8

UNCTAD. (2003). *Environmental Goods: Identifying Items of Export Interest to Developing Countries*. UNCTAD.

You, W. H., Zhu, H. M., Yu, K., & Peng, C. (2015). Democracy, financial openness, and global carbon dioxide emissions: Heterogeneity across existing emission levels. *World Development*, *66*, 189–207. doi:10.1016/j.worlddev.2014.08.013

Zugravu-Soilita, N. (2009). *Croissance, commerce, IDE et leur impact sur l'environnement: cas de l'Europe Centrale et Orientale et de la Communauté des Etats Indépendants* (Doctoral dissertation). Université Paris 1.

ENDNOTES

1 See Mankiw et al. (1992).

2 This factor is considered crucial to the development and the growth of income in comparison with the geographical and international trade factor (Sachs (2003)).

3 Theoretical analysis of Easterly & Levine (2003).

4 For the labor force expressed as a function of the labor force, we identified negative and significant signs coefficients at the 1% level. These results are consistent with those identified in Zugravu-Soilita (2009). This surprising result may justify the idea that job creation in some countries is carried out in the public sector (administrative tasks in the public service) outside production activities in the industrial sector and services generating added value. Thus, the labor factor in some cases decreases the level of national income. This result can also be explained by "moonlighting" in some economies. The state in this case cannot levy taxes. These illegal activities lead to lower state revenues and hinder the implementation of tax reform, resulting in a decline in the level of national income.

5 Following the method of Zugravu-Soilita (2009).

6 Following the method of Zugravu-Soilita (2009). For the calculation method, we only presented the example of the end-of-pipe products (EOP).

7 Following the method of Zugravu-Soilita (2009). For the calculation method, we only presented the example of the end-of-pipe products (EOP).

8 Organisation de coopération et de développement économiques. (2001). Mesurer la productivité: mesurer la croissance de la productivité par secteur et pour l'ensemble de l'économie: manuel de l'OCDE. Organisation de coopération et de développement économiques.

APPENDIX

Table 6. Definition of variables

Variables	Definition	Sources
$OECO_2$	CO_2 emissions (Metric tons per person) during the period 1995 – 2014.	*Energy Information Administration (2016)*
	For the year 2015, we relied on CO_2 emissions series in million tons after dividing each value by each country's population. (data is not available for some countries).	*Created and calculated by the author, based on the data of the BP Statistical Review of World Energy2016)*
Total population	**Variable of the total population used to calculate CO_2 emissions per capita for the year 2015.**	*World Bank (2016)*
GDP	GDP at constant prices ($US constant 2005).	*World Bank (2016), (WDI, (2016)) United Nation Statistics Division (2016) (National Accounts)*
TradeINT _ OA / A	Trade Intensity in OA product groups, Class A. It is defined by the share of exports and imports of OA products group in the overall exports and imports of goods.	*Created and calculated by the author, based on the data presented in UNComtrade Database (2016).*
TradeINT_EOP/OA	Trade Intensity in end-of-pipe products appearing on the OA list. It is defined as a function of the share of total exports and imports of EOP products in the overall exports and imports of goods.	*Created and calculated by the author, based on the data presented in UNComtrade Database (2016).*
TradeINT_CTP/OA	Trade Intensity in clean technologies and products of the OA list. It is defined as a function of the share of exports and imports of "green" or clean technologies and products in the overall exports and imports of goods.	*Created and calculated by the author, based on the data presented in UNComtrade Database (2016).*
Trade_ICT	Trade in Information and Communication Technologies. This variable is calculated as a function of the sum of ICT goods imports. (as a percentage of the total imports of goods) and ICT goods exports (as a percentage of the total exports of goods).	*Created and calculated by the author, based on the data presented in World Bank (2016), (WDI, (2016))*
Trade_SERV	Trade in services as a percentage of GDP. This variable is defined by imports and exports as a percentage of GDP (in US $).	*World Bank (2016), (WDI, (2016))*
GNP_p	Gross national income per capita, PPP (current international $).	*World Bank (2016), (WDI, (2016))))*

continued on following page

Table 6. Continued

Variables	Definition	Sources
ISEP	Index of the severity of environmental policy. This index varies from « 0 » (low degree of environmental policy stringency) to « 6 » (high degree of environmental policy stringency) and it takes into consideration a set of 14 measures and instruments focusing on the quality of the environment and pollution	*OECD, (2016)*
	In the case of developing countries, the index is expressed as « CPIA policy and institutions for environmental sustainability rating". The classification varies between "0" (low) and "6" (high).	*World Bank (2016), (WDI, (2016))*
GFCF	**Gross Fixed Capital Formation ($US constant 2005). This variable is used to calculate the capital stock.**	*World Bank (2016)*
K	The capital stock is calculated according to gross fixed capital ($US constant 2005). Method of calculation: we used a depreciation rate of 5%, fixed for most types of installations and equipments [8]. *Capital stock = fixed capital (of the year (t)) + capital stock created (of the year (t-1)) - (5% of the capital stock created (of the year (t-1)).*	*Created and calculated by the author, based on the data presented in World Bank (2016), (WDI, (2016))*
L	The labor factor is expressed as a function of the total working population.	*World Bank (2016), (WDI, (2016))*
K/L	Relative endowments in capital and labor. This variable is expressed as the quotient of the division of capital stock by labor (working population).	*Created and calculated by the author.*
Trade	The intensity of international openness is expressed as a function of The sum of exports and imports divided by the GDP (constant).	*World Bank (2016)); World Trade Organisation (2016), (WTO)* ▪ *UNCTAD Statistics Database (2016).* ▪ *Created and calculated by the author.*
Feed_Index	Democracy is measured by the civil liberty index and is calculated by a simple average of the political rights and the civil liberty index of each country (DP + IL) / 2.	*Created and calculated by the author, based on the data presented in Freedom House Organisation (2016)*
POLITstab	Indicator of political stability	*Kaufmann et al. (2006) Worldwide Governance Indicators (WGI), (2016)*
IC_imp	The concentration index on imports.	*UNCTAD Statistics Database (2016).*
IC_exp	The concentration index on export.	*UNCTAD Statistics Database (2016).*
IDiv_imp	The import diversification index.	*UNCTAD Statistics Database (2016).*
IDiv_exp	The export diversification index.	*UNCTAD Statistics Database (2016).*

Table 7. List of countries

Developing countries	
Low-income countries	*Middle income countries*
bangladesh	Kyrgyz Republic
Benign	Albania
Burkina Faso	Armenia
Burundi	Azerbaijan
Cambodia	Bolivia
Comoros	Bosnia and Herzegovina
Ethiopia	Bhutan
Gambia	Cameroon
Guinea	Ivory Coast
Kenya	Georgia
Madagascar	Ghana
Malawi	Grenade
Mali	Guyana
Mozambique	Honduras
Nepal	India
Niger	kiribati
Uganda	Lesotho
Central African Republic	moldova
Rwanda	Mongolia
Tanzania	Nigeria
Zimbabwe	Pakistan
Republic of Congo	tonga
Nicaragua	vanuatu
Dominican Republic	Viet Nam
St. LUCIA	Yemen
Saint Vincent and the Grenadines	Zambia
Sudan	samoa
Sri Lanka	Senegal
Togo	Sao Tome and Principe

Related Readings

To continue IGI Global's long-standing tradition of advancing innovation through emerging research, please find below a compiled list of recommended IGI Global book chapters and journal articles in the areas of waste management, environmental sustainability, and nanotechnology. These related readings will provide additional information and guidance to further enrich your knowledge and assist you with your own research.

Affam, A. C., & Chaudhuri, M. (2019). Comparative Study of Advance Oxidation Processes for Treatment of Pesticide Wastewater. In H. Aziz & S. Abu Amr (Eds.), *Advanced Oxidation Processes (AOPs) in Water and Wastewater Treatment* (pp. 261–323). Hershey, PA: IGI Global. doi:10.4018/978-1-5225-5766-1.ch011

Ahmedi, F., Ahmedi, L., O'Flynn, B., Kurti, A., Tahirsylaj, S., Bytyçi, E., ... Salihu, A. (2018). InWaterSense: An Intelligent Wireless Sensor Network for Monitoring Surface Water Quality to a River in Kosovo. *International Journal of Agricultural and Environmental Information Systems*, *9*(1), 39–61. doi:10.4018/IJAEIS.2018010103

Akash, P. V. M., Babu, N., & Navneet, (2018). Bioremediation of Environmental Pollutants. In V. Pathak, & Navneet (Eds.), Handbook of Research on Microbial Tools for Environmental Waste Management (pp. 80-104). Hershey, PA: IGI Global. doi:10.4018/978-1-5225-3540-9.ch005

Akbar, N. A., Matsin, Z. B., & Ramli, S. F. (2019). Groundwater Treatment via Ozonation. In H. Aziz & S. Abu Amr (Eds.), *Advanced Oxidation Processes (AOPs) in Water and Wastewater Treatment* (pp. 200–225). Hershey, PA: IGI Global. doi:10.4018/978-1-5225-5766-1.ch009

Ali, D. A., Palaniandy, P., & Feroz, S. (2019). Advanced Oxidation Processes (AOPs) to Treat the Petroleum Wastewater. In H. Aziz & S. Abu Amr (Eds.), *Advanced Oxidation Processes (AOPs) in Water and Wastewater Treatment* (pp. 99–122). Hershey, PA: IGI Global. doi:10.4018/978-1-5225-5766-1.ch005

Amorim, C. L., Moreira, I. S., Duque, A. F., van Loosdrecht, M. C., & Castro, P. M. (2017). Aerobic Granular Sludge: Treatment of Wastewaters Containing Toxic Compounds. In Á. Val del Río, J. Campos Gómez, & A. Mosquera Corral (Eds.), *Technologies for the Treatment and Recovery of Nutrients from Industrial Wastewater* (pp. 231–263). Hershey, PA: IGI Global. doi:10.4018/978-1-5225-1037-6.ch009

Anderson, S., & Rahman, P. K. (2018). Bioprocessing Requirements for Bioethanol: Sugarcane vs. Sugarcane Bagasse. In V. Pathak, & Navneet (Eds.), Handbook of Research on Microbial Tools for Environmental Waste Management (pp. 48-56). Hershey, PA: IGI Global. doi:10.4018/978-1-5225-3540-9.ch003

Ara, T., Bashir, R., Chisti, H., & Rangreez, T. A. (2019). Wastewater Pollution From the Industries. In A. Hussain & S. Ahmed (Eds.), *Advanced Treatment Techniques for Industrial Wastewater* (pp. 98–113). Hershey, PA: IGI Global. doi:10.4018/978-1-5225-5754-8.ch007

Aris, A., Jusoh, M. N., & Wahab, N. S. (2019). Applications of Advanced Oxidation Processes in Palm Oil Mill Effluent Treatment. In H. Aziz & S. Abu Amr (Eds.), *Advanced Oxidation Processes (AOPs) in Water and Wastewater Treatment* (pp. 123–149). Hershey, PA: IGI Global. doi:10.4018/978-1-5225-5766-1.ch006

Aziz, N. A., & Palaniandy, P. (2019). Photocatalysis (TiO2/Solar) in Water and Wastewater Treatment. In H. Aziz & S. Abu Amr (Eds.), *Advanced Oxidation Processes (AOPs) in Water and Wastewater Treatment* (pp. 171–199). Hershey, PA: IGI Global. doi:10.4018/978-1-5225-5766-1.ch008

Babu, N., & Pathak, V. M. (2018). Biosorption of Heavy Metals: Biological Approach to Control the Industrial Waste. In V. Pathak, & Navneet (Eds.), Handbook of Research on Microbial Tools for Environmental Waste Management (pp. 270-281). Hershey, PA: IGI Global. doi:10.4018/978-1-5225-3540-9.ch013

Bagheri, S., & Julkapli, N. M. (2017). Biomass-Derived Activated Carbon: Synthesis, Functionalized, and Photocatalysis Application. In T. Saleh (Ed.), *Advanced Nanomaterials for Water Engineering, Treatment, and Hydraulics* (pp. 162–199). Hershey, PA: IGI Global. doi:10.4018/978-1-5225-2136-5.ch007

Banerjee, P., Mukhopadhayay, A., & Das, P. (2016). Advances in Bioremediation for Removal of Toxic Dye from Different Streams of Wastewater. In A. Rathoure & V. Dhatwalia (Eds.), *Toxicity and Waste Management Using Bioremediation* (pp. 266–278). Hershey, PA: IGI Global. doi:10.4018/978-1-4666-9734-8.ch013

Basak, B., & Dey, A. (2016). Bioremediation Approaches for Recalcitrant Pollutants: Potentiality, Successes and Limitation. In A. Rathoure & V. Dhatwalia (Eds.), *Toxicity and Waste Management Using Bioremediation* (pp. 178–197). Hershey, PA: IGI Global. doi:10.4018/978-1-4666-9734-8.ch009

Basheer, T., & Umesh, M. (2018). Valorization of Tannery Solid Waste Materials Using Microbial Techniques: Microbes in Tannery Solid Waste Management. In V. Pathak, & Navneet (Eds.), Handbook of Research on Microbial Tools for Environmental Waste Management (pp. 127-145). Hershey, PA: IGI Global. doi:10.4018/978-1-5225-3540-9.ch007

Becker, J. A., & Stefanakis, A. I. (2016). Pharmaceuticals and Personal Care Products as Emerging Water Contaminants. In A. McKeown & G. Bugyi (Eds.), *Impact of Water Pollution on Human Health and Environmental Sustainability* (pp. 81–100). Hershey, PA: IGI Global. doi:10.4018/978-1-4666-9559-7.ch004

Belmonte, M., Fajardo, C., Toledo-Alarcón, J. B., Heredia, D. V., Jorquera, L., Méndez, R., ... Ruiz-Filippi, G. (2017). Autotrophic Denitrification Processes. In Á. Val del Río, J. Campos Gómez, & A. Mosquera Corral (Eds.), *Technologies for the Treatment and Recovery of Nutrients from Industrial Wastewater* (pp. 147–173). Hershey, PA: IGI Global. doi:10.4018/978-1-5225-1037-6.ch006

Bender, S. R. (2016). Floodplain Infrastructure and the Toxic Tide. In A. McKeown & G. Bugyi (Eds.), *Impact of Water Pollution on Human Health and Environmental Sustainability* (pp. 150–173). Hershey, PA: IGI Global. doi:10.4018/978-1-4666-9559-7.ch007

Benjamin, S. R., de Lima, F., & Rathoure, A. K. (2016). Genetically Engineered Microorganisms for Bioremediation Processes: GEMs for Bioremediaton. In A. Rathoure & V. Dhatwalia (Eds.), *Toxicity and Waste Management Using Bioremediation* (pp. 113–140). Hershey, PA: IGI Global. doi:10.4018/978-1-4666-9734-8.ch006

Bhakta, J. N. (2016). Microbial Response against Metal Toxicity. In A. Rathoure & V. Dhatwalia (Eds.), *Toxicity and Waste Management Using Bioremediation* (pp. 75–96). Hershey, PA: IGI Global. doi:10.4018/978-1-4666-9734-8.ch004

Bhatt, P. (2018). Insilico Tools to Study the Bioremediation in Microorganisms. In V. Pathak, & Navneet (Eds.), Handbook of Research on Microbial Tools for Environmental Waste Management (pp. 389-395). Hershey, PA: IGI Global. doi:10.4018/978-1-5225-3540-9.ch018

Bogatu, C., & Leszczynska, D. (2016). Transformation of Nanomaterials in Environment: Surface Activation of SWCNTs during Disinfection of Water with Chlorine. *Journal of Nanotoxicology and Nanomedicine*, *1*(1), 45–57. doi:10.4018/JNN.2016010104

Bonmatí-Blasi, A., Cerrillo-Moreno, M., & Riau-Arenas, V. (2017). Systems Based on Physical-Chemical Processes: Nutrient Recovery for Cycle Closure. In Á. Val del Río, J. Campos Gómez, & A. Mosquera Corral (Eds.), *Technologies for the Treatment and Recovery of Nutrients from Industrial Wastewater* (pp. 43–75). Hershey, PA: IGI Global. doi:10.4018/978-1-5225-1037-6.ch003

Cema, G., & Sochacki, A. (2017). Treatment of Landfill Leachate by Anammox Process. In Á. Val del Río, J. Campos Gómez, & A. Mosquera Corral (Eds.), *Technologies for the Treatment and Recovery of Nutrients from Industrial Wastewater* (pp. 290–311). Hershey, PA: IGI Global. doi:10.4018/978-1-5225-1037-6.ch011

Cervantes, F. J., Cuervo-López, F., & Hernández, J. G. (2017). Fundamentals of the Biological Processes for Nitrogen Removal. In Á. Val del Río, J. Campos Gómez, & A. Mosquera Corral (Eds.), *Technologies for the Treatment and Recovery of Nutrients from Industrial Wastewater* (pp. 112–146). Hershey, PA: IGI Global. doi:10.4018/978-1-5225-1037-6.ch005

Chaudhry, S., & Garg, S. (2019). Industrial Wastewater Pollution and Advanced Treatment Techniques. In A. Hussain & S. Ahmed (Eds.), *Advanced Treatment Techniques for Industrial Wastewater* (pp. 74–97). Hershey, PA: IGI Global. doi:10.4018/978-1-5225-5754-8.ch006

Chaurasia, P. K., Bharati, S. L., & Mani, A. (2018). Enzymatic Treatment of Petroleum-Based Hydrocarbons. In V. Pathak, & Navneet (Eds.), Handbook of Research on Microbial Tools for Environmental Waste Management (pp. 396-408). Hershey, PA: IGI Global. doi:10.4018/978-1-5225-3540-9.ch019

Cook, C., & Gude, V. G. (2017). Characteristics of Chitosan Nanoparticles for Water and Wastewater Treatment: Chitosan for Water Treatment. In T. Saleh (Ed.), *Advanced Nanomaterials for Water Engineering, Treatment, and Hydraulics* (pp. 223–261). Hershey, PA: IGI Global. doi:10.4018/978-1-5225-2136-5.ch009

Cuker, B., & Bugyi, G. (2016). The Fundamentals of Water and Natural Waters. In A. McKeown & G. Bugyi (Eds.), *Impact of Water Pollution on Human Health and Environmental Sustainability* (pp. 1–28). Hershey, PA: IGI Global. doi:10.4018/978-1-4666-9559-7.ch001

de la Varga, D., Soto, M., Arias, C. A., van Oirschot, D., Kilian, R., Pascual, A., & Álvarez, J. A. (2017). Constructed Wetlands for Industrial Wastewater Treatment and Removal of Nutrients. In Á. Val del Río, J. Campos Gómez, & A. Mosquera Corral (Eds.), *Technologies for the Treatment and Recovery of Nutrients from Industrial Wastewater* (pp. 202–230). Hershey, PA: IGI Global. doi:10.4018/978-1-5225-1037-6.ch008

Dhatwalia, V. K., & Nanda, M. (2016). Biodegradation of Phenol: Mechanisms and Applications. In A. Rathoure & V. Dhatwalia (Eds.), *Toxicity and Waste Management Using Bioremediation* (pp. 198–214). Hershey, PA: IGI Global. doi:10.4018/978-1-4666-9734-8.ch010

Di Paola, N., Spanò, R., Caldarelli, A., & Vona, R. (2016). Fighting Ecomafias: The Role of Biotech Networks in Achieving Sustainability. In A. Rathoure & V. Dhatwalia (Eds.), *Toxicity and Waste Management Using Bioremediation* (pp. 295–311). Hershey, PA: IGI Global. doi:10.4018/978-1-4666-9734-8.ch015

Dima, J. B., & Zaritzky, N. (2017). Performance of Chitosan Micro/Nanoparticles to Remove Hexavalent Chromium From Residual Water. In T. Saleh (Ed.), *Advanced Nanomaterials for Water Engineering, Treatment, and Hydraulics* (pp. 262–288). Hershey, PA: IGI Global. doi:10.4018/978-1-5225-2136-5.ch010

Dubey, S. K., Hussain, A., & Ajnavi, S. (2019). Environmental Recycling System (ERS): An Emerging Approach to Solid Waste Management. In A. Hussain & S. Ahmed (Eds.), *Advanced Treatment Techniques for Industrial Wastewater* (pp. 225–237). Hershey, PA: IGI Global. doi:10.4018/978-1-5225-5754-8.ch013

Farraji, H. (2018). Phytoremediation of Nitrogen and Phosphorus in Municipal Wastewater by Cyperus alternifolius Planted Constructed Wetland. In V. Pathak, & Navneet (Eds.), Handbook of Research on Microbial Tools for Environmental Waste Management (pp. 146-163). Hershey, PA: IGI Global. doi:10.4018/978-1-5225-3540-9.ch008

Farraji, H., Zaman, N. Q., Vakili, M., & Faraji, H. (2016). Overpopulation and Sustainable Waste Management. *International Journal of Sustainable Economies Management*, *5*(3), 13–36. doi:10.4018/IJSEM.2016070102

Fra-Vázquez, A., & Valenzuela-Heredia, D. (2017). Biogeochemical Cycles of Nitrogen and Phosphorus: Implications of Anthropogenic Activities. In Á. Val del Río, J. Campos Gómez, & A. Mosquera Corral (Eds.), *Technologies for the Treatment and Recovery of Nutrients from Industrial Wastewater* (pp. 1–20). Hershey, PA: IGI Global. doi:10.4018/978-1-5225-1037-6.ch001

Gatt, K. (2016). Youths' Social Traits in Water Management as a Precursor for Good Water Governance. *International Journal of Information Systems and Social Change*, *7*(3), 16–26. doi:10.4018/IJISSC.2016070102

Gupta, C., & Prakash, D. (2016). Novel Bioremediation Methods in Waste Management: Novel Bioremediation Methods. In A. Rathoure & V. Dhatwalia (Eds.), *Toxicity and Waste Management Using Bioremediation* (pp. 141–157). Hershey, PA: IGI Global. doi:10.4018/978-1-4666-9734-8.ch007

Hassan, S. H., & Halim, A. A. (2019). Water Quality Legislation and Regulation. In H. Aziz & S. Abu Amr (Eds.), *Advanced Oxidation Processes (AOPs) in Water and Wastewater Treatment* (pp. 30–45). Hershey, PA: IGI Global. doi:10.4018/978-1-5225-5766-1.ch002

Hilles, A. H., Abu Amr, S. S., Aziz, H. A., & Bashir, M. J. (2019). Advanced Oxidation Processes for Water and Wastewater Treatment: An Introduction. In H. Aziz & S. Abu Amr (Eds.), *Advanced Oxidation Processes (AOPs) in Water and Wastewater Treatment* (pp. 46–69). Hershey, PA: IGI Global. doi:10.4018/978-1-5225-5766-1.ch003

Hivrale, A. U., Pawar, P. K., Rane, N. R., & Govindwar, S. P. (2016). Application of Genomics and Proteomics in Bioremediation. In A. Rathoure & V. Dhatwalia (Eds.), *Toxicity and Waste Management Using Bioremediation* (pp. 97–112). Hershey, PA: IGI Global. doi:10.4018/978-1-4666-9734-8.ch005

Hussain, A., Bhattacharya, A., & Ahmed, A. (2019). Plastic Waste Pollution and Its Management in India: A Review. In A. Hussain & S. Ahmed (Eds.), *Advanced Treatment Techniques for Industrial Wastewater* (pp. 62–73). Hershey, PA: IGI Global. doi:10.4018/978-1-5225-5754-8.ch005

Ibrahim, N., Zainal, S. F., & Aziz, H. A. (2019). Application of UV-Based Advanced Oxidation Processes in Water and Wastewater Treatment. In H. Aziz & S. Abu Amr (Eds.), *Advanced Oxidation Processes (AOPs) in Water and Wastewater Treatment* (pp. 384–414). Hershey, PA: IGI Global. doi:10.4018/978-1-5225-5766-1.ch014

Jagdale, S., & Chabukswar, A. (2016). Phyto-Remediation: Using Plants to Clean Up Soils: Phyto-Remediation. In A. Rathoure & V. Dhatwalia (Eds.), *Toxicity and Waste Management Using Bioremediation* (pp. 215–235). Hershey, PA: IGI Global. doi:10.4018/978-1-4666-9734-8.ch011

Kamaruddin, M. A. (2019). Advanced Oxidation Processes (AOPs) in Landfill Leachate Treatment. In H. Aziz & S. Abu Amr (Eds.), *Advanced Oxidation Processes (AOPs) in Water and Wastewater Treatment* (pp. 355–383). Hershey, PA: IGI Global. doi:10.4018/978-1-5225-5766-1.ch013

Kaur, A. (2017). Applications of Nanomaterials for Water Treatment: A Future Avenue. In T. Saleh (Ed.), *Advanced Nanomaterials for Water Engineering, Treatment, and Hydraulics* (pp. 289–304). Hershey, PA: IGI Global. doi:10.4018/978-1-5225-2136-5.ch011

Kehinde, F. O., Abu Amr, S. S., & Aziz, H. A. (2019). Application of Persulfate in Textile Wastewater Treatment. In H. Aziz & S. Abu Amr (Eds.), *Advanced Oxidation Processes (AOPs) in Water and Wastewater Treatment* (pp. 70–98). Hershey, PA: IGI Global. doi:10.4018/978-1-5225-5766-1.ch004

Khelfi, A. (2018). Sources of Air Pollution. In V. Pathak, & Navneet (Eds.), Handbook of Research on Microbial Tools for Environmental Waste Management (pp. 220-258). Hershey, PA: IGI Global. doi:10.4018/978-1-5225-3540-9.ch011

Khelfi, A. (2018). Sources of Soil Pollution. In V. Pathak, & Navneet (Eds.), Handbook of Research on Microbial Tools for Environmental Waste Management (pp. 190-218). Hershey, PA: IGI Global. doi:10.4018/978-1-5225-3540-9.ch010

Khelfi, A. (2019). Sources of Groundwater Pollution. In A. Hussain & S. Ahmed (Eds.), *Advanced Treatment Techniques for Industrial Wastewater* (pp. 177–210). Hershey, PA: IGI Global. doi:10.4018/978-1-5225-5754-8.ch011

Koli, S. K., & Hussain, A. (2019). Status of Electronic Waste Management in India: A Review. In A. Hussain & S. Ahmed (Eds.), *Advanced Treatment Techniques for Industrial Wastewater* (pp. 238–250). Hershey, PA: IGI Global. doi:10.4018/978-1-5225-5754-8.ch014

Kulkarni, A. G., & Vedamurthy, A. B. (2018). Cellulolytic Microorganisms: A Review. In V. Pathak, & Navneet (Eds.), Handbook of Research on Microbial Tools for Environmental Waste Management (pp. 34-47). Hershey, PA: IGI Global. doi:10.4018/978-1-5225-3540-9.ch002

Kumar, P. S., & C., F. C. (2019). Soil Bioremediation Techniques. In A. Hussain, & S. Ahmed (Eds.), *Advanced Treatment Techniques for Industrial Wastewater* (pp. 35-50). Hershey, PA: IGI Global. doi:10.4018/978-1-5225-5754-8.ch003

Kumar, P. S., & Saravanan, A. (2019). Water Pollutants and Their Removal Techniques. In A. Hussain & S. Ahmed (Eds.), *Advanced Treatment Techniques for Industrial Wastewater* (pp. 114–133). Hershey, PA: IGI Global. doi:10.4018/978-1-5225-5754-8.ch008

Kumar, P. S., & Yaashikaa, P. R. (2019). Agriculture Pollution. In A. Hussain & S. Ahmed (Eds.), *Advanced Treatment Techniques for Industrial Wastewater* (pp. 134–154). Hershey, PA: IGI Global. doi:10.4018/978-1-5225-5754-8.ch009

Laha, S. D., Dutta, K., & Kundu, P. P. (2018). Biodegradation of Low Density Polyethylene Films. In V. Pathak, & Navneet (Eds.), Handbook of Research on Microbial Tools for Environmental Waste Management (pp. 282-318). Hershey, PA: IGI Global. doi:10.4018/978-1-5225-3540-9.ch014

Manyuchi, M. M., Muzenda, E., & Mbohwa, C. (2018). Design and Development of a Sanitary Landfill for Low Income Countries for Optimal Waste Management. In V. Pathak, & Navneet (Eds.), Handbook of Research on Microbial Tools for Environmental Waste Management (pp. 373-388). Hershey, PA: IGI Global. doi:10.4018/978-1-5225-3540-9.ch017

Mbuh, M. J., Houser, P. R., & Heidari, A. (2016). Water Quality Estimation Using Combined Water Chemistry and Field Spectroscopy in the Shenandoah River, Virginia. *International Journal of Applied Geospatial Research, 7*(2), 14–37. doi:10.4018/ijagr.2016040102

McKeown, A. E. (2016). Nurses, Healthcare, and Environmental Pollution and Solutions: Breaking the Cycle of Harm. In A. McKeown & G. Bugyi (Eds.), *Impact of Water Pollution on Human Health and Environmental Sustainability* (pp. 174–197). Hershey, PA: IGI Global. doi:10.4018/978-1-4666-9559-7.ch008

Mobin, R., Chisti, H., Rangreez, T. A., Bashir, R., & Najar, A. A. (2019). Pesticidal Pollution. In A. Hussain & S. Ahmed (Eds.), *Advanced Treatment Techniques for Industrial Wastewater* (pp. 155–176). Hershey, PA: IGI Global. doi:10.4018/978-1-5225-5754-8.ch010

Mohamed, Z. (2016). Cyanobacterial Toxins in Water Sources and Their Impacts on Human Health. In A. McKeown & G. Bugyi (Eds.), *Impact of Water Pollution on Human Health and Environmental Sustainability* (pp. 120–149). Hershey, PA: IGI Global. doi:10.4018/978-1-4666-9559-7.ch006

Mojiri, A., Ziyang, L., Hui, W., & Gholami, A. (2019). Concentrated Landfill Leachate Treatment by Electro-Ozonation. In H. Aziz & S. Abu Amr (Eds.), *Advanced Oxidation Processes (AOPs) in Water and Wastewater Treatment* (pp. 150–170). Hershey, PA: IGI Global. doi:10.4018/978-1-5225-5766-1.ch007

Mujere, N. (2016). Water Hyacinth: Characteristics, Problems, Control Options, and Beneficial Uses. In A. McKeown & G. Bugyi (Eds.), *Impact of Water Pollution on Human Health and Environmental Sustainability* (pp. 343–361). Hershey, PA: IGI Global. doi:10.4018/978-1-4666-9559-7.ch015

Mujere, N., & Isidro, M. (2016). Impacts of Artisanal and Small-Scale Gold Mining on Water Quality in Mozambique and Zimbabwe. In A. McKeown & G. Bugyi (Eds.), *Impact of Water Pollution on Human Health and Environmental Sustainability* (pp. 101–119). Hershey, PA: IGI Global. doi:10.4018/978-1-4666-9559-7.ch005

Nazir, R. (2017). Advanced Nanomaterials for Water Engineering and Treatment: Nano-Metal Oxides and Their Nanocomposites. In T. Saleh (Ed.), *Advanced Nanomaterials for Water Engineering, Treatment, and Hydraulics* (pp. 84–126). Hershey, PA: IGI Global. doi:10.4018/978-1-5225-2136-5.ch005

Obeng, P. A., Oduro-Kwarteng, S., Keraita, B., Bregnhøj, H., Abaidoo, R. C., Awuah, E., & Konradsen, F. (2016). On-Site Sanitation Systems for Low-Income Countries: Technical Guidelines for Groundwater Pollution and Nuisance Control. In A. McKeown & G. Bugyi (Eds.), *Impact of Water Pollution on Human Health and Environmental Sustainability* (pp. 319–342). Hershey, PA: IGI Global. doi:10.4018/978-1-4666-9559-7.ch014

Olajubu, E. A., Ajayi, G., Oke, I., & Asahiah, F. O. (2017). An ANN Model for Predicting the Quantity of Lead and Cadmium Ions in Industrial Wastewater. *International Journal of Information Communication Technologies and Human Development, 9*(4), 32–44. doi:10.4018/IJICTHD.2017100103

Paris, R. S. (2016). Issues with Water Quality: How Do We Get Our Fellow Citizens to Care? In A. McKeown & G. Bugyi (Eds.), *Impact of Water Pollution on Human Health and Environmental Sustainability* (pp. 199–214). Hershey, PA: IGI Global. doi:10.4018/978-1-4666-9559-7.ch009

Pedemonte, D. C., Frison, N., Tayà, C., Ponsa, S., & Fatone, F. (2017). Chemical and Biological Processes for Nutrients Removal and Recovery. In Á. Val del Río, J. Campos Gómez, & A. Mosquera Corral (Eds.), *Technologies for the Treatment and Recovery of Nutrients from Industrial Wastewater* (pp. 76–111). Hershey, PA: IGI Global. doi:10.4018/978-1-5225-1037-6.ch004

Prabu, S. L., Suriyaprakash, T., Kandasamy, R., & Rathinasabapathy, T. (2016). Effective Waste Water Treatment and its Management. In A. Rathoure & V. Dhatwalia (Eds.), *Toxicity and Waste Management Using Bioremediation* (pp. 312–334). Hershey, PA: IGI Global. doi:10.4018/978-1-4666-9734-8.ch016

Priyanka, M., Kumar, D., Shankar, U., Yadav, A., & Yadav, K. (2018). Agricultural Waste Management for Bioethanol Production. In V. Pathak, & Navneet (Eds.), Handbook of Research on Microbial Tools for Environmental Waste Management (pp. 1-33). Hershey, PA: IGI Global. doi:10.4018/978-1-5225-3540-9.ch001

Rajasekaran, R. N., Puddoo, H., & Nguyenhuynh, T. N. (2019). An Overview of Treatment of Antibiotics Using Advanced Oxidation Process. In H. Aziz & S. Abu Amr (Eds.), *Advanced Oxidation Processes (AOPs) in Water and Wastewater Treatment* (pp. 226–260). Hershey, PA: IGI Global. doi:10.4018/978-1-5225-5766-1.ch010

Rangreez, T. A., Mobin, R., Chisti, H., Bashir, R., & Ara, T. (2019). Advanced Water Treatment Systems and Their Applications. In A. Hussain & S. Ahmed (Eds.), *Advanced Treatment Techniques for Industrial Wastewater* (pp. 1–18). Hershey, PA: IGI Global. doi:10.4018/978-1-5225-5754-8.ch001

Rastogi, M., & Nandal, M. (2018). Composting an Emerging Technology for Solid Waste Management in India. In V. Pathak, & Navneet (Eds.), Handbook of Research on Microbial Tools for Environmental Waste Management (pp. 105-126). Hershey, PA: IGI Global. doi:10.4018/978-1-5225-3540-9.ch006

Rathoure, A. K. (2016). Heavy Metal Pollution and its Management: Bioremediation of Heavy Metal. In A. Rathoure & V. Dhatwalia (Eds.), *Toxicity and Waste Management Using Bioremediation* (pp. 27–50). Hershey, PA: IGI Global. doi:10.4018/978-1-4666-9734-8.ch002

Sadegh, H., & Ali, G. A. (2019). Potential Applications of Nanomaterials in Wastewater Treatment: Nanoadsorbents Performance. In A. Hussain & S. Ahmed (Eds.), *Advanced Treatment Techniques for Industrial Wastewater* (pp. 51–61). Hershey, PA: IGI Global. doi:10.4018/978-1-5225-5754-8.ch004

Said, S., Hussain, A., & Sharma, G. (2018). Water Quality Mapping of Yamuna River Stretch Passing Through Delhi State Using High Resolution Geoeye-2 Imagery. *International Journal of Applied Geospatial Research*, *9*(4), 23–35. doi:10.4018/IJAGR.2018100102

Saleh, T. A. (2017). An Overview of Nanomaterials for Water Technology. In T. Saleh (Ed.), *Advanced Nanomaterials for Water Engineering, Treatment, and Hydraulics* (pp. 1–12). Hershey, PA: IGI Global. doi:10.4018/978-1-5225-2136-5.ch001

Saleh, T. A., Adio, S. O., Parthasarathy, P., & Danmaliki, G. I. (2017). Scientific Insights Into Modified and Non-Modified Biomaterials for Sorption of Heavy Metals From Water. In T. Saleh (Ed.), *Advanced Nanomaterials for Water Engineering, Treatment, and Hydraulics* (pp. 13–39). Hershey, PA: IGI Global. doi:10.4018/978-1-5225-2136-5.ch002

Saleh, T. A., & Danmaliki, G. I. (2017). Polymer Consumption, Environmental Concerns, Possible Disposal Options, and Recycling for Water Treatment. In T. Saleh (Ed.), *Advanced Nanomaterials for Water Engineering, Treatment, and Hydraulics* (pp. 200–222). Hershey, PA: IGI Global. doi:10.4018/978-1-5225-2136-5.ch008

Saleh, T. A., Majeed, S., Nayak, A., & Bhushan, B. (2017). Principles and Advantages of Microwave-Assisted Methods for the Synthesis of Nanomaterials for Water Purification. In T. Saleh (Ed.), *Advanced Nanomaterials for Water Engineering, Treatment, and Hydraulics* (pp. 40–57). Hershey, PA: IGI Global. doi:10.4018/978-1-5225-2136-5.ch003

Satpathy, R. (2018). Microbial Bioreactor Systems for Dehalogenation of Organic Pollutants. In V. Pathak, & Navneet (Eds.), Handbook of Research on Microbial Tools for Environmental Waste Management (pp. 319-340). Hershey, PA: IGI Global. doi:10.4018/978-1-5225-3540-9.ch015

Shankara, S., Dupadahalli, K., Vijayakumar, M. H., & Gaddad, S. M. (2016). Decolorization of Direct Blue: 14 Dye by Thermoalkalophilic Aerobic Bacillus sp. In A. Rathoure & V. Dhatwalia (Eds.), *Toxicity and Waste Management Using Bioremediation* (pp. 279–294). Hershey, PA: IGI Global. doi:10.4018/978-1-4666-9734-8.ch014

Sharma, A., Kaur, M., Katnoria, J. K., & Nagpal, A. K. (2016). Heavy Metal Pollution: A Global Pollutant of Rising Concern. In A. Rathoure & V. Dhatwalia (Eds.), *Toxicity and Waste Management Using Bioremediation* (pp. 1–26). Hershey, PA: IGI Global. doi:10.4018/978-1-4666-9734-8.ch001

Shikuku, V. O., & Achieng', G. O. (2019). Occurrence and Fate of Selected Heavy Metals in a Conventional Municipal Wastewater Treatment Plant in Kisumu City, Kenya: A Case Study. In A. Hussain & S. Ahmed (Eds.), *Advanced Treatment Techniques for Industrial Wastewater* (pp. 211–224). Hershey, PA: IGI Global. doi:10.4018/978-1-5225-5754-8.ch012

Shikuku, V. O., Kowenje, C. O., & Nyairo, W. N. (2017). Fundamentals and Sources of Magnetic Nanocomposites and Their Sorption Properties. In T. Saleh (Ed.), *Advanced Nanomaterials for Water Engineering, Treatment, and Hydraulics* (pp. 58–83). Hershey, PA: IGI Global. doi:10.4018/978-1-5225-2136-5.ch004

Shikuku, V. O., Nyairo, W. N., & Kowenje, C. O. (2019). Preparation and Application of Biochars for Organic and Microbial Control in Wastewater Treatment Regimes. In A. Hussain & S. Ahmed (Eds.), *Advanced Treatment Techniques for Industrial Wastewater* (pp. 19–34). Hershey, PA: IGI Global. doi:10.4018/978-1-5225-5754-8.ch002

Shrivas, K., Ghosale, A., & Maji, P. (2017). Advanced Nanomaterials for the Removal of Chemical Substances and Microbes From Contaminated and Waste Water. In T. Saleh (Ed.), *Advanced Nanomaterials for Water Engineering, Treatment, and Hydraulics* (pp. 127–161). Hershey, PA: IGI Global. doi:10.4018/978-1-5225-2136-5.ch006

Simmonds, J., Gómez, J. A., & Ledezma, A. (2018). Statistical and Data Mining Techniques for Understanding Water Quality Profiles in a Mining-Affected River Basin. *International Journal of Agricultural and Environmental Information Systems, 9*(2), 1–19. doi:10.4018/IJAEIS.2018040101

Singh, R., & Chauhan, M. (2016). Effective Management of Agro-Industrial Residues as Composting in Mushroom Industry and Utilization of Spent Mushroom Substrate for Bioremediation. In A. Rathoure & V. Dhatwalia (Eds.), *Toxicity and Waste Management Using Bioremediation* (pp. 158–177). Hershey, PA: IGI Global. doi:10.4018/978-1-4666-9734-8.ch008

Smith, M. C. (2016). Design of the Education Campaign: Creating a Motivated, Environmentally Concerned Public through Improved Scientific Literacy. In A. McKeown & G. Bugyi (Eds.), *Impact of Water Pollution on Human Health and Environmental Sustainability* (pp. 215–239). Hershey, PA: IGI Global. doi:10.4018/978-1-4666-9559-7.ch010

Starkl, M., Brunner, N., Hauser, A. W., Feil, M., & Kasan, H. (2018). Addressing Sustainability of Sanitation Systems: Can it be Standardized? *International Journal of Standardization Research, 16*(1), 39–51. doi:10.4018/IJSR.2018010103

Stefanakis, A. I. (2016). Constructed Wetlands: Description and Benefits of an Eco-Tech Water Treatment System. In A. McKeown & G. Bugyi (Eds.), *Impact of Water Pollution on Human Health and Environmental Sustainability* (pp. 281–303). Hershey, PA: IGI Global. doi:10.4018/978-1-4666-9559-7.ch012

Stefanakis, A. I., & Becker, J. A. (2016). A Review of Emerging Contaminants in Water: Classification, Sources, and Potential Risks. In A. McKeown & G. Bugyi (Eds.), *Impact of Water Pollution on Human Health and Environmental Sustainability* (pp. 55–80). Hershey, PA: IGI Global. doi:10.4018/978-1-4666-9559-7.ch003

Stratigea, A., Leka, A., & Panagiotopoulou, M. (2017). In Search of Indicators for Assessing Smart and Sustainable Cities and Communities' Performance. *International Journal of E-Planning Research*, *6*(1), 43–73. doi:10.4018/IJEPR.2017010103

Tajarudin, H. A., Makhtar, M. M., Azmi, M. S., Zainuddin, N. I., & Ali, D. H. (2019). Introduction to Water and Wastewater Treatment. In H. Aziz & S. Abu Amr (Eds.), *Advanced Oxidation Processes (AOPs) in Water and Wastewater Treatment* (pp. 1–29). Hershey, PA: IGI Global. doi:10.4018/978-1-5225-5766-1.ch001

Tischler, D., Qi, J., Ngo, A. C., & Schlömann, M. (2018). Microbial Degradation of Azo Dyes: The Role of Azoreductase to Initiate Degradation. In V. Pathak, & Navneet (Eds.), Handbook of Research on Microbial Tools for Environmental Waste Management (pp. 341-371). Hershey, PA: IGI Global. doi:10.4018/978-1-5225-3540-9.ch016

Turcea, V. C., & Mihai, M. M. (2019). The Impact of CO2 Emissions From Waste on Human Health: A Consequence of Globalization. *International Journal of Sustainable Economies Management*, *8*(1), 17–28. doi:10.4018/IJSEM.2019010102

Umesh, M., & Basheer, T. (2018). Microbe Mediated Bioconversion of Fruit Waste Into Value Added Products: Microbes in Fruit Waste Management. In V. Pathak, & Navneet (Eds.), Handbook of Research on Microbial Tools for Environmental Waste Management (pp. 57-78). Hershey, PA: IGI Global. doi:10.4018/978-1-5225-3540-9.ch004

Val del Río, Á., Fernández, P. C., Gómez, J. L., & Mosquera-Corral, A. (2017). Nutrients Pollution in Water Bodies: Related Legislation in Europe and the United States. In Á. Val del Río, J. Campos Gómez, & A. Mosquera Corral (Eds.), *Technologies for the Treatment and Recovery of Nutrients from Industrial Wastewater* (pp. 21–42). Hershey, PA: IGI Global. doi:10.4018/978-1-5225-1037-6.ch002

Val del Río, Á., Fuentes, A. P., Giustinianovich, E. A., Gomez, J. L., & Mosquera-Corral, A. (2017). Anammox Process: Technologies and Application to Industrial Effluents. In Á. Val del Río, J. Campos Gómez, & A. Mosquera Corral (Eds.), *Technologies for the Treatment and Recovery of Nutrients from Industrial Wastewater* (pp. 264–289). Hershey, PA: IGI Global. doi:10.4018/978-1-5225-1037-6.ch010

Valverde-Pérez, B., Flores-Alsina, X., Vangsgaard, A. K., & Mauricio-Iglesias, M. (2017). Modelling and Control of Nitrogen and Phosphorus Removing Systems. In Á. Val del Río, J. Campos Gómez, & A. Mosquera Corral (Eds.), *Technologies for the Treatment and Recovery of Nutrients from Industrial Wastewater* (pp. 174–201). Hershey, PA: IGI Global. doi:10.4018/978-1-5225-1037-6.ch007

Verma, A., & Shukla, P. K. (2016). A Prospective Study on Emerging Role of Phytoremediation by Endophytic Microorganisms. In A. Rathoure & V. Dhatwalia (Eds.), *Toxicity and Waste Management Using Bioremediation* (pp. 236–265). Hershey, PA: IGI Global. doi:10.4018/978-1-4666-9734-8.ch012

Vipond, E., & Rahman, P. K. (2018). Anaerobic Digestion Enhancement With Microbial Electrolysis Cells: Is Biomethane Production the Direction to Go for Commercialization? In V. Pathak, & Navneet (Eds.), Handbook of Research on Microbial Tools for Environmental Waste Management (pp. 259-269). Hershey, PA: IGI Global. doi:10.4018/978-1-5225-3540-9.ch012

Voulvoulis, N., & Georges, K. (2016). Industrial and Agricultural Sources and Pathways of Aquatic Pollution. In A. McKeown & G. Bugyi (Eds.), *Impact of Water Pollution on Human Health and Environmental Sustainability* (pp. 29–54). Hershey, PA: IGI Global. doi:10.4018/978-1-4666-9559-7.ch002

Wahla, V., & Shukla, S. (2018). Role of Microorganisms in Bioremediation of Pesticides. In V. Pathak, & Navneet (Eds.), Handbook of Research on Microbial Tools for Environmental Waste Management (pp. 164-189). Hershey, PA: IGI Global. doi:10.4018/978-1-5225-3540-9.ch009

Willems, R. C., & MacDonald, S. A. (2016). Access to Safe Drinking Water: A Low Income Sustainable Water Purification System. In A. McKeown & G. Bugyi (Eds.), *Impact of Water Pollution on Human Health and Environmental Sustainability* (pp. 304–318). Hershey, PA: IGI Global. doi:10.4018/978-1-4666-9559-7.ch013

Zakaria, S. N. (2019). Ozonation With Catalyst in Landfill Leachate Treatment. In H. Aziz & S. Abu Amr (Eds.), *Advanced Oxidation Processes (AOPs) in Water and Wastewater Treatment* (pp. 324–354). Hershey, PA: IGI Global. doi:10.4018/978-1-5225-5766-1.ch012

Zaman, A., Das, P., & Banerjee, P. (2016). Biosorption of Dye Molecules. In A. Rathoure & V. Dhatwalia (Eds.), *Toxicity and Waste Management Using Bioremediation* (pp. 51–74). Hershey, PA: IGI Global. doi:10.4018/978-1-4666-9734-8.ch003

Ziemba, R., Cuker, B. E., Stein, J., Meuninck, R., & Wan, J. A. (2016). Health Professionals Can Protect Water Quality: Tools for Educators, Advocates, and Practitioners. In A. McKeown & G. Bugyi (Eds.), *Impact of Water Pollution on Human Health and Environmental Sustainability* (pp. 240–280). Hershey, PA: IGI Global. doi:10.4018/978-1-4666-9559-7.ch011

About the Contributors

Sang-Bing Tsai is currently a Professor at University of Electronic Science and Technology of China Zhongshan Institute, Capital University of Economics and Business and Wuyi University. He is both Technology Management and Business Management PhD. He has published many research articles in SCI / SSCI journals, including International Journal of Production Research, International Journal of Web Services Research, Advances in Mechanical Engineering, IEEE Transactions on Systems Man Cybernetics-Systems, Plos One, Environment and Planning B-Planning & Design, Sustainability, Renewable & Sustainable Energy Reviews,...etc. His recent research interests in Computer Science, Information Management, Big Data, Sustainability, Green Operation and Applied Mathematics. He has well over 100 published peer-reviewed journal articles. Dr. Tsai is will be the Co-Editor-in-Chief for the Journal of Organizational and End User Computing. He is currently the Associate Editor for Journal of Global Information Management, and Associate Editor for International Journal of Digital Crime and Forensics, as well as serving on the editorial boards of 10 other journals.

* * *

Muhmmad Haseeb Ahmad did his doctorate from University of Hohenheim, Stuttgart Germany in 2016.After completion of his doctoral studies, he came back to Pakistan and joined National Institute of Food Science and Technology, University of Agriculture, Faisalabad. He served that institute for one and half year. Currently he is working in Institute of Home and Food Sciences, Government College University, Faisalabad as an Assistant Professor. He is the author of 15 impacted research articles along with three book chapters.

Sélima Ben Zineb is a professor-researcher with a doctorate in economics at the Faculty of Economics and Management of Tunis El Manar (FSEGT). She is currently a member of the research laboratory Financial Development and Innovation (DEFI) at ESSEC in Tunis.

Jishnu Bhattacharyya is educated in management science at Indian Institute of Management Kozhikode and Indian Institute of Information Technology and Management (ABV-IIITM) Gwalior. His engineering education was under the umbrella of the Maulana Abul Kalam Azad University of Technology, West Bengal (Formerly known as WBUT). He is interested broadly in Marketing Science and particularly interested in Consumer churn behaviour, Digital Marketing, Marketing Analytics, Strategy Management, Social Marketing, and Sustainability Marketing. To his credit, he has some international Conference Papers and published journal papers.

Savita Budhwar is currently working as an Assistant Professor in the Department of Nutrition Biology, Central University of Haryana, India. Her research interests include but not limited to the value addition of cereals and pulses, waste management and public health.

Bin Ding is a lecturer in Human Resource Management at the Xi'an Jiaotong-Liverpool University in China. Prior to that, she worked at the Robert H. Smith Business School at University of Maryland, College Park as a postdoc. Dr. Ding obtained her Ph.D. in Management at the University of Lausanne in Switzerland. Her research interests include but not limited to service management, organizational climate, public health, management information systems, and leadership.

Anum Ishaq is a PhD from National Institute of Food Science and Technology (NIFSAT), University of Agriculture, Faisalabad (UAF), Pakistan and currently working as a lecturer at Department of Allied Health Sciences, Superior University Lahore, Pakistan. She has published 12 research and review articles in various journals of national and international repute with impact factor of about 15.137, 4 book chapters and 34 scientific abstracts as proceedings. Her research projects are mainly focused on improving Food Safety and Quality by using various chemical, thermal, non-thermal, physical and biological interventions. During her M.Sc research, she worked under the title of "Inhibition of surface spoilage bacteria in broiler meat by utilizing organic acids" while in PhD, she worked under the title of "Employing multiple hurdle approach for microbial decontamination and quality enhancement of meat". At Purdue University (USA), she worked on the use of bacteriophages as biological intervention for inhibiting the growth of pathogens from beef surface.

Song Jiang received the M.S. degrees in mine engineering from the School of Management, Xi'an University of Architecture and Technology, Shaanxi, China in 2012, He is currently pursuing a Ph.D. at School of Management, Xi'an University of Architecture and Technology, Shaanxi, China. He is the author of two books, more than 20 articles. His research interests include mining system engineering, big data and mining management. He has a strong interest in the exploration of interdisciplinary fields of computers and mine.

Ben Krishna is currently pursuing PhD in Information Technology and Systems at Indian Institute of Management, Kozhikode. He has completed his Executive Post-graduate Diploma from Indian Institute of Management, Kozhikode and Bachelor of Technology in Electronics and Communication Engineering from Government Engineering College, Thrissur. Also, he has industry Experience as Business Analyst with 6 years in business analysis, market research and servicing clients in Healthcare, Retail and Banking domain and latest worked as a Techno marketing Consultant for a Healthcare client. His interest areas include the behavioral side of Information technology. He has published and presented papers in International conferences.

Soumyadeep Kundu is currently a doctoral candidate in the Quantitative Methods & Operations Management area at IIM Kozhikode. He graduated with a B.Tech in Electronics & Electrical Engineering from Kalinga Institute of Industrial Technology, Bhubaneswar finishing in the top percentile of his class. His research interest lie in the retail sector particularly assortment planning and fast fashion retail, urban mobility, micro-mobility, consumer information search and behavioral operations. He is also involved in case writing for class room teaching purposes and occasionally writes pieces for his publication 'The OPs Guy' on Medium.

Junaid Ahmad Malik has completed his Ph.D in Zoology (Wildlife) in 2015. He is working as a Lecturer in Zoology at Govt. Degree College, Bijbehara, Anantnag. Dr. Malik has published many research articles and technical papers in many International and National reputed Journals. He has also authored several books and book chapters of reputed publishers. He is in the editorial board and a regular reviewer of several journals of reasonable reputation.

Mukesh Narmetta is currently pursuing his Ph.D. from the Indian Institute of Management, Kozhikode specializing in Information Systems area. His research interests are to understand the impact of technology on society, Human Computer Interaction & e-learning. He has three and a half years of Corporate experience as Business Analyst & Developer with an under graduation in Electronics and Communication Engineering from JNT University, Hyderabad.

Ubaid ur Rahman is a PhD from National Institute of Food Science and Technology (NIFSAT), University of Agriculture, Faisalabad (UAF), Pakistan and currently working as Assistant Professor at School of Food and Agricultural Sciences, University of Management and Technology, Lahore, Pakistan. He has published 18 research and review articles in various journals of national and international repute with impact factor of about 44.139, 11 book chapters and 59 scientific abstracts as proceedings. Presently, he is also paying his services as Editorial Board Member of SCIREA Journal of Food, Member of Food Safety Advisory Board, Cambridge Scholars Publishing, Member of Food Science and Technology Advisory Board, Cambridge Scholars Publishing and Potential Reviewer of several International Journals. His research projects are mainly focused on addressing the national issues of Food Safety, Quality and Authenticity. During PhD, he worked under the title of "Assessment of meat quality and microbial safety through molecular & spectroscopic techniques coupled with chemometrics". In these projects, he worked on the development of rapid analytical techniques for detection of meat spoilage, quality, presence of adulterants, contamination and Halal authenticity. Additionally, he hs also worked in numerous other research projects of my supervisor at University of Agriculture, Faisalabad (Pakistan). The key projects include "Rapid & Non-destructive evaluation of food quality and safety using spectroscopy" funded by Higher Education Commission (HEC), Pakistan, "Assessment of microbial quality of raw meat & development of rapid technique to determine the microbial load on the surface of fresh meat" funded by Higher Education Commission (HEC), Pakistan and "Rapid and Non-destructive Methods for Detection of Fraud Ground Meat to Ensure Food Safety for Consumers" under PERIDOT Research Program 2017, funded by Higher Education Commission (HEC), Pakistan. At Purdue University (USA), he worked on Exploring the potential of Laser Optical Scattering Technology for rapid detection of food-borne pathogenic bacteria in isolated and mixed cultures, and Identification of internalization routes of pathogens in cucumbers through DNA identification.

Olea Roy has a Bachelor of Science in Economics from St. Xavier's College, Kolkata and a Master of Arts in Economics from Jadavpur University. Her research interests reside in Behavioral Economics and Public Policy.

Qamar Abbas Syed completed his doctoral studies in food science and technology from Univerisitat Autonoma de Barcelona, Spain. He has been working in dairy and beverage industries in research and development projects. Presently he is working as Assistant Professor of Food Science at University of Agriculture, Faisalabad, Pakistan. His expertise include processing of dairy, fruit juices and carbonated beverages. He has successfully completed research projects related to high pressure processing and ultraviolet treatments of dairy and juices.

Wenqing Wu is currently the Associate Professor at College of Management and Economics Tianjin University of China. He is Technology Economy and Management PhD. He has published many research articles in SCI / SSCI journals, including Physica A: Statistical Mechanics, Journal of Artificial Societies and Social Simulation, Scientometrics, Sustainability, etc. Dr. Wu serves more than 10 other journals, such as Physica A: Statistical Mechanics, Internet Research, IEEE Access, Plos one. His recent research interests in Social System Simulation, Green operation, Sustainability.

Index

www.ingramcontent.com/pod-product-compliance
Lightning Source LLC
LaVergne TN
LVHW061535290926
849162LV00004B/798